Applied Intelligent Decision Making in Machine Learning

Computational Intelligence in Engineering Problem Solving
Series Editor: Nilanjan Dey

Computational Intelligence (CI) can be framed as a heterogeneous domain that harmonizes and coordinates several technologies, such as probabilistic reasoning, artificial life, multi-agent systems, neuro-computing, fuzzy systems, and evolutionary algorithms. Integrating several disciplines, such as Machine Learning (ML), Artificial Intelligence (AI), Decision Support Systems (DSS), and Database Management Systems (DBMS), increases the CI power and impact in several engineering applications. This series provides an excellent forum for discussing the characteristics of CI systems in engineering. It emphasizes the development of CI techniques and their role as well as the state-of-the-art solutions in different real-world engineering applications. The series is aimed at researchers, academics, scientists, engineers, and other professionals who are involved in the new techniques of CI, which, including artificial fuzzy logic and neural networks, are presented for biomedical image processing, power systems, and reactor applications.

Applied Machine Learning for Smart Data Analysis
Nilanjan Dey, Sanjeev Wagh, Parikshit N. Mahalle, Mohd. Shafi Pathan

IoT Security Paradigms and Applications
Research and Practices
Sudhir Kumar Sharma, Bharat Bhushan, Narayan C. Debnath

Applied Intelligent Decision Making in Machine Learning
Himansu Das, Jitendra Kumar Rout, Suresh Chandra Moharana, Nilanjan Dey

For more information about this series, please visit: https://www.crcpress.com/Computational-Intelligence-in-Engineering-Problem-Solving/book-series/CIEPS

Applied Intelligent Decision Making in Machine Learning

Edited by

Himansu Das
Jitendra Kumar Rout
Suresh Chandra Moharana
Nilanjan Dey

CRC Press
Taylor & Francis Group
Boca Raton London New York

CRC Press is an imprint of the
Taylor & Francis Group, an **Informa** business

First edition published 2021
by CRC Press
6000 Broken Sound Parkway NW, Suite 300, Boca Raton, FL 33487-2742

and by CRC Press
2 Park Square, Milton Park, Abingdon, Oxon, OX14 4RN

© 2021 Taylor & Francis Group, LLC

CRC Press is an imprint of Taylor & Francis Group, an Informa business

No claim to original U.S. Government works

**Visit the Taylor & Francis Web site at
http://www.taylorandfrancis.com**

**and the CRC Press Web site at
http://www.crcpress.com**

ISBN: 978-0-367-50336-9 (hbk)
ISBN: 978-1-003-04954-8 (ebk)

Typeset in Times
by SPi Global, India

Contents

Preface

Decision making is the process of selection of best choice made logically from the accessible options. Decision making analyzes the large volume of data in a particular field and is a very difficult task to do manually. A technique must be used to find out the possible outcome of each option available. It also determines the option that is best at a particular moment for an effective decision-making process. Machine learning uses a set of mathematical models and algorithms for the construction of a decision-making system that gradually improves its performance on specific tasks. It is also based on a model that develops intelligent, adaptive, and hybrid tools and methods for solving complex decision-making problems. Artificial intelligence has already been recognized as a standard tool for the decision-making process. Despite several such developments, many findings still need to be addressed due to the advancement of technologies such as artificial intelligence, machine learning, and information technology. The computer system should analyze dynamically each option as well as alternatives to make an appropriate intelligent decision. This book will also specifically focus on applied intelligent decision-making fields that perform various tasks effectively for predictive analysis, classification, clustering, feature selection, feature reduction, recommendations, information retrieval, and pattern recognition in a very limited amount of time with high performance.

The objective of this edited book is to reach academic scientists, researchers, data scientists, and scholars to exchange and share experiences and research outcomes in various domains to develop cost-effective, intelligent, and adaptive models that handle various challenges in the field of decision making and to help researchers carry it forward to a next level. The book incorporates the advances of machine intelligent techniques in the decision-making process and its applications. It also provides a premier interdisciplinary platform for scientists, researchers, practitioners, and educators to share their thoughts in the context of recent innovations, trends, developments, practical challenges, and advancements in the field of data mining, machine learning, soft computing, and decision science. It addresses recent developments and applications in machine learning. It also focuses on the usefulness of applied intelligent techniques in the decision-making process in several ways. To address these objectives, the book includes 12 chapters contributed by several authors from across the globe.

Chapter 1 concerns the mining of streaming data for the extraction of knowledge for decision making and provides efficient algorithms to do this. In Chapter 2, the authors use two agricultural applications to demonstrate decoded machine learning methods. In Chapter 3, the authors design a hybrid model for optical character recognition to effectively recognize the character classes. Chapter 4 analyzes hybrid computational approaches, like the hybrid biogeography-based optimization technique for landslide susceptibility that models by considering the different causal factors. In Chapter 5, the author suggests an article according to a content-based approach, by embedding a seed document (query) into a linear space to find its nearest neighbors as candidates, and these candidates are re-ranked using a model that discriminates between cited and uncited publications. Chapter 6 focuses on the analysis of air quality data, based on air pollutant concentration information in India. In Chapter 7, the authors address the improvised binary shuffled frog leaping algorithm and meta-heuristic methodology for biomarker gene determination. In Chapter 8, the authors present a method which identifies multiword entities and categorizes them into the appropriate type of concept. The approach uses the notion of matrix-based multi-pattern matching to generate the frequency of patterns. Chapter 9 applies different machine-learning techniques such as a decision tree, a random forest, extreme gradient boosting, and support vector regression to model Australia's carbon emissions. Chapter 10 analyzes daily forex data in terms of 1 USD to five different currencies (Australian Dollar, Canadian Dollar, British Pound, Indian Rupee, and Japanese Yen); over 1600 samples are predicted accurately and efficiently with the use of a deep learning technique. Chapter 11 addresses a teaching-learning-based optimization for feature selection (FS) to find the optimal features by updating the weak features with the strong features. Chapter 12 uses a dehazing technique for contrast-limited adaptive histogram equalization and a guided filter to regain the original quality of the image.

The topics brought together in this book are unique and are also based on the unpublished work of the contributing authors. We have tried to communicate all the novel experiments and directions that have been made in the area of machine learning, decision science, and its diversified applications. We believe this book provides a reference text for larger audiences such as practitioners, system architects, researchers, and developers.

Himansu Das
KIIT Deemed to be University, Bhubaneswar, Odisha, India

Jitendra Kumar Rout
KIIT Deemed to be University, Bhubaneswar, Odisha, India

Suresh Chandra Moharana
KIIT Deemed to be University, Bhubaneswar, Odisha, India

Nilanjan Dey
Techno India College of Technology, Rajarhat, Kolkata, India

Notes on the Editors and Contributors

Himansu Das works as Assistant Professor in the School of Computer Engineering, Kalinga Institute of Industrial Technology (KIIT) Deemed to be University, Bhubaneswar, India. He received a B.Tech degree from the Institute of Technical Education and Research, India and an M.Tech degree in Computer Science and Engineering from the National Institute of Science and Technology, India. He has published several research papers in various international journals and has presented at conferences. He has also edited several books published by IGI Global, Springer, and Elsevier. He has also served on many journals and conferences as editorial or reviewer board member. He is proficient in the field of Computer Science Engineering and served as an organizing chair, a publicity chair, and acted as a member of the technical program committees of many national and international conferences. He is also associated with various educational and research societies, such as IET, IACSIT, ISTE, UACEE, CSI, IAENG, and ISCA. His research interests include the field of Data Mining, Soft Computing, and Machine Learning. He has more than ten years teaching and research experience in various engineering colleges.

Jitendra Kumar Rout is Assistant Professor in the School of Computer Engineering, KIIT Deemed to be University, Bhubaneswar, India. He completed his Masters and PhD at the National Institute of Technology, Rourkela, India in 2013 and 2017 respectively. Prior to KIIT University he was Lecturer in various engineering colleges, such as GITA and TITE Bhubaneswar. He is a life member of the Odisha IT Society (OITS) and has been actively associated with conferences such as ICIT (one of the oldest in Odisha). He is also a life member of IEI, and a member of IEEE, ACM, IAENG, and UACEE. He has published in IEEE and with Springer. His main research interests include Data Analytics, Machine Learning, NLP, Privacy in Social Networks, and Big Data.

Suresh Chandra Moharana is Assistant Professor at the School of Computer Engineering at KIIT Deemed to be University. He obtained a Bachelor of Engineering degree in Computer Science and Engineering in 2003, and a Master in Technology degree in Information Security in 2011. Currently, he is pursuing his PhD work in Computer Science and Engineering in the same university. His research efforts are focused on Cloud Computing, Virtualization, and Data Analytics. At the research level, he has published around ten articles in reputed international journals.

Nilanjan Dey is Assistant Professor in the Department of Information Technology at Techno India College of Technology (under Techno India Group), Kolkata, India. He is a Visiting Fellow of the University of Reading, UK and Visiting Professor of Duy Tan University, Vietnam. He was an honorary Visiting Scientist at Global

Biomedical Technologies Inc., CA, USA (2012–2015). He is also a research scientist at the Laboratory of Applied Mathematical Modeling in Human Physiology, Territorial Organization of Scientific and Engineering Unions, Bulgaria. He is Associate Researcher at the Laboratoire RIADI, University of Manouba, Tunisia and is a scientific member of Politécnica of Porto. He was awarded his PhD from Jadavpur University in 2015. In addition, he was awarded for being one of the top 10 most published academics in the field of Computer Science in India (2015–2017) during the Faculty Research Awards organized by Careers 360 in New Delhi, India. Before he joined Techno India, he was Assistant Professor of JIS College of Engineering and Bengal College of Engineering and Technology. He has authored/edited more than 50 books, and published more than 300 papers. His h-index is 32 with more than 4900 citations. He is the Editor-in-Chief of the *International Journal of Ambient Computing and Intelligence* and the *International Journal of Rough Sets and Data Analysis*. He is Co-Editor-in-Chief of the *International Journal of Synthetic Emotions* and the *International Journal of Natural Computing Research*. He is the Series Co-Editor of Springer Tracts in Nature-Inspired Computing (STNIC), Springer, the Series Co-Editor of Advances in Ubiquitous Sensing Applications for Healthcare (AUSAH), Elsevier, and the Series Editor of Computational Intelligence in Engineering Problem Solving and Intelligent Signal Processing and Data Analysis, CRC Press (FOCUS/Brief Series) and Advances in Geospatial Technologies (AGT) Book Series, (IGI Global), US, and serves as an editorial board member of several international journals, including the *International Journal of Image Mining*, and as Associated Editor of IEEE Access (SCI-Indexed) and the *International Journal of Information Technology*. His main research interests include Medical Imaging, Machine Learning, Computer Aided Diagnosis as well as Data Mining. He has been on program committees of over 50 international conferences, has organized five workshops, and acted as a program co-chair and/or advisory chair of more than ten international conferences. He has given more than 40 invited lectures in ten countries, including many invited plenary/keynote talks at international conferences, such as ITITS2017 (China), TIMEC2017 (Egypt), SOFA2018 (Romania), and BioCom2018 (UK).

1 Data Stream Mining for Big Data

Chandresh Kumar Maurya

CONTENTS

1.1 INTRODUCTION

There are two kinds of learning, based on how the data is processed. In *batch* learning, the data is processed in chunks and often *offline*. Another type of learning is called *online* learning, usually performed on *streaming* data. Another name for online learning is *incremental* learning. Our focus in this chapter will be incremental learning on streaming data.

A data stream is characterized by certain properties according to Gama (2010):

- Unbounded size:
  - Transient (lasts for only few seconds or minutes);
  - Single-pass over data;
  - Only summaries can be stored;
  - Real-time processing (in-memory).
- Data streams are not static:
  - Incremental/decremental updates;
  - Concept drifts.
- Temporal order may be important.

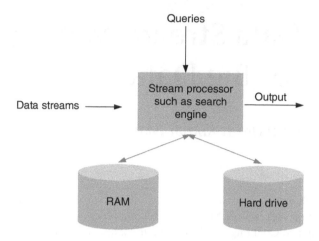

FIGURE 1.1   General stream processing model.

Traditional algorithms developed for offline processing of the data are not suitable for streaming data due to these issues. There are a few models that are incremental such as K-nearest neighbors (KNN) and naive Bayes. But, still, these models cannot cope with all the issues present in streaming data. For example, if there is a concept drift present in the data, a separate concept-drift detection and alleviation technique needs to be developed. Many researchers have proposed various models for streaming data which tackles some of the issues above. We will look at some of these techniques in detail in the coming sections.

As shown in Figure 1.1, a stream processor takes as input a stream of data such as 0, 1, 1, 0, 1, …. The processor can sample from the stream and process it in main memory, such as when answering a query, and the result of the processing can be dumped into the back-end, such as the hard drive, for downstream tasks if required. Most of the algorithms presented in the subsequent sections will be based on this generic model.

Data stream mining finds application in several domains such as answering user queries over the web, sensor data (Das et al., 2019; Dey et al., 2019), analyzing network packets, patient health monitoring, and surveillance systems to name a few. Because of the vast application area of the field, I will present some selected case studies in the field at the end of the chapter.

The rest of the chapter is as follows. Research issues related to stream mining are discussed in Section 1.2. Section 1.3 elucidates the simple problems in streaming data such as filtering and counting. Sampling from a data stream is explained in Section 1.4, followed by concept drift detection in Section 1.5. Finally, the chapter ends with a discussion and an open problem in Section 1.6.

1.2 RESEARCH ISSUES IN DATA STREAM MINING

Knowledge discovery from a data stream is a challenging task due to reasons mentioned in the previous section. Most challenging is the unbounded size of the stream, low memory, and fast arrival rate. As a result, techniques proposed in the literature

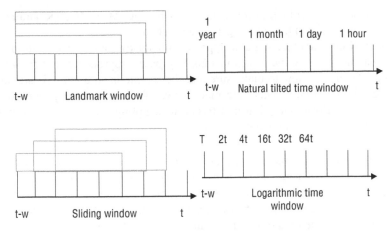

FIGURE 1.2 (Left) Sequence-based windows. Top figure is a landmark window and bottom figure is a sliding window (used in packet transmission). (Right) Timestamp-based windows. Top figure is a natural tilted window and bottom figure is a logarithmic time window.

rely on ways to summarize the data stream in some way. The following are the techniques often employed to handle data streams.

1. **Approximate query processing** (AQP) is described according to the following rules (Chaudhuri et al., 2017): (a) a query language should be generic enough to retrieve all the relevant information, (b) should have a low error rate and high accuracy, (c) the work should be saved at runtime, and (d) extra resources should be made available for pre-computation. These dimensions of an AQP scheme are not independent and much of the previous literature makes specific choices along the above four dimensions. AQP can be broadly classified into two main categories: (a) online aggregation, which is a sampling in an online manner and using samples to answer queries, and (b) offline synopses generation, which is based on some statistics of the data and which then answers queries (Li and Li, 2018).

2. **Sliding window query processing** is a model of handling a data stream with the assumption that recent items are more important than the older ones. There are two types of sliding window based on how they are defined.
   - **Sequence-based sliding window** which is defined according to the number of observations that determine the window size. These are further divided into *landmark sliding window* and *moving sliding window*, as shown in Figure 1.2 (left).
   - **Timestamp-based sliding window:** In this case, the sliding window is determined based on the number of observations made in a given time interval and where window-width is determined based on the time. They are further categorized into *natural-titled time window* and *logarithmic time window*, as shown in Figure 1.2 (right).

   As we shall see, some algorithms based on the sliding window guarantee an error rate of approximation to any desired accuracy ϵ. One of the most popular algorithms in this category is the DGIM method (Rajaraman and Ullman, 2011) for counting 1s in a bit stream.

3. **Sampling** is another popular approach for making an approximation of the streaming data. Some issues in sampling-based approaches are: (a) when to sample and (b) how to make sure that the sample is an unbiased estimator of the original stream. There are many ways of doing sampling. For example, random uniform sampling and reservoir sampling are some of the popular approaches used on the data stream.

In the forthcoming sections, I present algorithms for some basic tasks in data stream mining. These are filtering and counting in the data stream, sampling from the data stream, and concept-drift detection in data streams. For other tasks, such as classification, clustering, frequent pattern mining, and novelty detection in data stream mining, the interested reader is referred to Gama (2010) and Aggarwal (2007).

1.3 FILTERING AND COUNTING IN A DATA STREAM

In this section, we discuss algorithms for filtering items which satisfy a property and others which count in a data stream.

1.3.1 BLOOM FILTERS

Bloom filters are one of the most popular approaches for filtering an item from a data stream. Applications of bloom filters include checking if a user-id is available for sign-up, keeping track of web-pages visited by a user, and recent email addresses used to send messages. A bloom filter consists of three things (Rajaraman and Ullman, 2011).

- A bit array of length n initialized to 0.
- A number of hash functions $h_1, h_2, ..., h_k$. Each hash function maps "keys" to bit-array value by setting the bit.
- A set of m key values.

We will see an application of the bloom filter used to sign-up at various platforms for finding user-ids. A bloom filter gives guarantees of the form: If a user-id is available, it's 100% sure of its answer. On the other hand, if a user-id is not available, it gives a probabilistic answer to the query. Let us see this in action. A bloom filter is shown in Figure 1.3. In this figure, there are two hash functions h_1 and h_2 which can map a given string to the bit-array. Suppose the strings "cow" and "chicken" are mapped to $(6, 3)$ and $(2, 1)$ locations. Bits at these locations are set to 1 in the bit-array. Now suppose a query for string "pigeon" is issued. If the hash functions generate hashes 0 and 3, then we see that bit at location 0 is not set. This means that "pigeon" can be used as a user-id. However, if the hashes generated were, say, $(2, 3)$, then it means that "pigeon" may be already taken as a user-id which the bloom filter is not sure

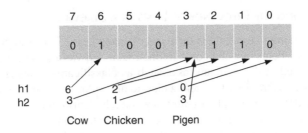

FIGURE 1.3   Hashing items in a bit-array. h_1 and h_2 are hash functions.

about. The reason is that some other strings might have set the same bit locations as "pigeon" in this scenario and there is no way to be sure who set that bit. For this reason, it's impossible to delete the user-id using a bit-array. To solve this problem, another variation of the bloom filter which uses *counts* of the "strings" to a location is used. Such bloom filters are called *count-bloom filters*. Some important points to note while designing bloom filters are the following:

1. The size of the bloom filter is very important. The smaller the bloom filter, the larger the false positive (FP) and vice versa.
2. The larger the number of hash functions, the quicker the bloom filter fills, as well as slow filter. Too few hash functions results in too many FPs, unless all are *good* hash functions.

A visualization of the FP rate with the number of hash functions is given in Figure 1.4. As we can see, as the number of hash functions grows, the FP rate declines sharply. For a more in-depth analysis of the bloom filter, the reader is referred to Rajaraman and Ullman (2011).

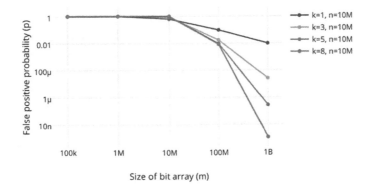

FIGURE 1.4   Effect of the number of hash functions on the false positive rate. Image source: Bhat (2016).

1.3.2 COUNTING THE FREQUENCY OF ITEMS IN A STREAM

Frequency counting in a stream is the basis of many tasks. For example, if we know the frequency of items, we can compute the mean and variances of them which can be further utilized in downstream tasks such as classification, clustering, and computing similarity. Here, I will present a classic algorithm for counting the frequency of items in a data stream called *count-min* sketch (Cormode and Muthukrishnan, 2004). A count-min algorithm can be seen as an extension of the bloom filter discussed above. The key idea of the algorithm is to map the domain of the stream from $\{1,...,n\}$ to $\{1,...,d\}$ using hash functions.

Hash values are stored in a two-dimensional array called *sketches* of width $w = \lceil e \ / \ \epsilon \rceil$ and height $h = \left\lceil \log \dfrac{1}{\delta} \right\rceil$, where (ϵ, δ) denote the accuracy in the approximation and the confidence level, respectively.

An example of a count-min algorithm is given in Figure 1.5. To understand the algorithm, let us take an example. Imagine the data stream looks like:

$$...A, B, A, C, A, A, C,... \tag{1.1}$$

Initially, the sketch is filled with 0s. The algorithm takes each letter at a time and applies hash functions. For example, letter "A" hashes to columns $(0, 1, 3, 4, 2)$ by the hash functions in the sketch. Corresponding entries in the sketch are incremented by 1. Similarly, all other letters are hashed. During query time, the same hash function is applied to produce the entries in the sketch. The frequency of the letter is given by taking the minimum of these entries. The count-min sketch is guaranteed to never give false negatives. That means the estimate is either one of true values or over-counts. Over-counts happen when two or more hash functions map at least two letters at exactly the same locations. That is why the choice of a good random hash function is required for the working of the algorithm. For a more in-depth discussion, the reader can consult the original article (Cormode and Muthukrishnan, 2004).

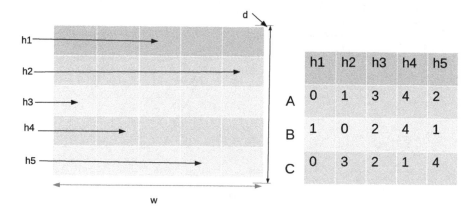

FIGURE 1.5 Count-min sketch. Hash functions map the domain of the stream to a limited range (left). An example of the hash functions mapping letters to integers.

1.3.3 COUNT UNIQUE ITEMS IN A DATA STREAM

Counting the number of unique items in the stream presents an interesting problem. Formally, this problem can be posed as follows. Suppose the domain of the random variable is $\{0, 1, ..., n\}$ and the stream looks like:

$$0, 1, 0, 0, 0, 1, 1, 2, 3, 3...$$

Here $n = 3$. A trivial solution is to store all new items in a hash map whose size grows like $O(n)$. However, we want a solution which runs only in $O(\log(n))$. One elegant solution is given by Flajolet and Martin (FM) (1985). FM is given in Algorithm 1.1. Let us work through the algorithm step-by-step.

- We first initialize a BITMAP to all 0s.
- Then for each item x in the stream, compute a hash value for it. Call it $h(x)$.
- The binary representation of the hash value is $bin(h(x))$.
- The function ρ gives the position of the least significant set-bit. This also occurs if ρ returns 0 for the 0 bit.
- Finally set the BITMAP[i] = 1.

Algorithm 1.1: Flajolet and Martin (FM) Algorithm

Input: stream M
Output: Cardinality of M
1 initialization: BITMAP OF L bits initialized to 0.;
2 **for** *each x in M* **do**
  - Calculate hash function $h(x)$.;
  - get binary representation of hash output, call it $bin(h(x))$.;
  - Calculate the index i such that $i = \rho(bin(h(x)))$, where $\rho()$ is such that it outputs the **position** of the least-significant set bit,i.e., position of 1.;
  - set BITMAP[i] = 1.;

3 **end**
4 Let R denote the largest index i such that BITMAP[i] = 1;
5 Cardinality of M is $2^R/\phi$ where $\phi \approx 0.77351$;

The above steps are repeated until the end of the stream. Let R denote the index of the most significant set-bit. Then, the number of unique items is given by $2^R/\phi$ where $\phi \approx 0.77351$ is a correction factor.

The intuition behind the working of the FM algorithm is as follows. The FM algorithm uses a hash function that maps the strings in the stream to uniformly generate random integers in the range $[0, ..., 2^L - 1]$. Thus, we expect that:

- 1/2 of the numbers will have their binary representation end in 0 (divisible by 2).
- 1/4 of the numbers will have their binary representation end in 00 (divisible by 4).
- 1/8 of the numbers will have their binary representation end in 00 (divisible by 8).
- In general, $1/2^R$ of the numbers will have their binary representation end in 0^R.

Then, the number of unique strings will be approximately 2^R (because using n bits, we can represent 2^n integers). A worked out example of the FM algorithm is as follows.

Assume stream $M = \{\ldots, 1, 1, 2, 3, 1, 4, 2, \ldots\}$ and the hash function $h(x) = (2x + 1) \% 5$.

Then,

$$h(M) = \{3, 3, 0, 2, 3, 4, 0\}$$

and

$$\rho(h(M)) = \{0, 0, 0, 1, 0, 2, 0\}$$

and BITMAP looks like

$$\text{BITMAP} = 0 \,|\, 0 \,|\, 0 \,|\, 1 \,|\, 1 \,|\, 1$$

so that we see the largest integer has index $i = 2$. So $R = 2$ and the unique integers are $2^2/\phi = 5$ approximately (the exact answer is 5).

Exercise: Try finding the unique numbers in $M = \{5, 10, 15, 20, \ldots\}$ with any hash function. Extensions:loglog and *Hyper* log log algorithms are an improvement over the classic FM algorithm. The interested reader may consult Flajolet et al. (2007).

1.4 SAMPLING FROM DATA STREAMS

Due to the unbounded size of the data streams and low memory, we need some efficient ways to reduce the data footprint. In the literature, researchers usually use the following techniques to reduce the data load.

- Sampling.
- Sliding window.
- Histogram.
- Wavelets and Fourier transforms.

We have already seen some examples of sliding windows and they find application at several places such as the DGIM method (Rajaraman and Ullman, 2011) for counting 1s in a bit stream and so on. In this section, we are going to discuss a sampling method from data streams.

There are many ways to sample from a data stream. For example, we can use *random uniform sampling* which is the simplest algorithm. However, doing random uniform sampling in a data stream is not straightforward since all items in the stream are not available at once. Hence, there is another algorithm called *reservoir sampling* used to sample items from a data stream such that the probability of selecting an item is $1/n$ where n is the stream size.

The intuition behind reservoir sampling is as follows. Imagine a swimming pool which can have at most k people in it at any one time. Swimmers are coming continuously and any new swimmer can only enter the pool after a previous swimmer has come out. But, how do you decide which new swimmer to let in? One solution could be to toss a coin (biased) with probability $p = k/n$ where n is the number of swimmers who have arrived so far. If it comes up heads, let the swimmer in, otherwise deny entry. A reservoir sampling algorithm is given in Algorithm 1.2.

Algorithm 1.2: Reservoir Sampling Algorithm

---

   input S: stream of values, k: size of the reservoir
   :
   /* Create uniform sample of fixed size   */
1   Insert first k elements of S to the reservoir.
2   **for** $v \in S$ **do**
3   Let i be the index of v in S
4   Generate a random int $j \in [1, i]$
5   **if** $j < k$ **then**
6   Insert v into the reservoir
7   Delete an item from the reservoir at random
8   **end**
9 **end**

The algorithm follows the intuition as given above. At the beginning, k items in the stream are inserted into the reservoir. For the other incoming items, a random number is generated between 1 and i where i is the index of the items just arrived. If the random number thus generated lies in $[1, k]$, we insert the new item in the reservoir by deleting an item uniformly at random. Otherwise, discard the new item. The question that comes up is: What is the probability that the ith item will replace an old item?

Theorem 1.1

The probability that the new incoming item will replace the old one is given by $P(a_i \in R) = k/n$, where R denotes the reservoir, k the reservoir size, and n the stream size.

Proof: Note that the item i has probability k/i of being inserted into the reservoir R. (Why?) After that no other item should replace it otherwise it will be gone. Now, the next $(i + 1)$th item will not replace it, and has probability $\left(1 - \dfrac{1}{i+1}\right)$; similarly, the $(i + 2)$th item will not replace it and has probability $\left(1 - \dfrac{1}{i+2}\right)$; and so on. Since all of these probabilities are independent (?), we multiply all of them to get the desired result. $P\left(a_i \in R\right) = \dfrac{k}{i} \times \dfrac{i}{i+1} \times \dfrac{i+1}{i+2} \times \cdots, \dfrac{n-1}{n} = \dfrac{k}{n}$ □

Thus, we have proved that by using reservoir sampling we can obtain samples which are unbiased. Thus samples can be used for downstream tasks.

1.5 CONCEPT DRIFT DETECTION IN DATA STREAMS

Real-life data is continuously changing due to the dynamics of the complex environment. As a result, the assumption which most traditional learning algorithms make, namely that the data is independent and identically distributed, does not hold.

The problem is more severe in the case of data streams because of the high data rate and the non-availability of all the data points.

More formally, in data mining and machine learning, we try to find the function f that maps input x to output y.

$$y = f(x) \qquad (1.2)$$

Such a function is assumed to be *stationary*, i.e., the distribution generating the data is fixed (but unknown). But, real-life data is:

- Non-stationary.
- Evolving.

Therefore, we need methods that can detect and adapt to changes in the underlying function. The underlying function is called a *concept* and changes in the function is called *concept-drift*, also known as *change detection*, *covariate shift*, *dataset shift*, etc.

We should make a difference between *noise* and *concept-drift* here. A change in the underlying function is due to the noise that occurs when there are not sufficient examples present in the change region or they cannot form a new concept of their own. The main difficulty lies in determining the threshold when changes in the data form a new concept as opposed to remaining as noise. An example of concept-drift is shown in Figure 1.6. As we can see in the figure, there is a change in the variance of the signal; so it is a concept drift of signals.

1.5.1 CAUSES OF CONCEPT DRIFT

Often, the causes of the drift are unknown. However, most researchers try to find causes based on the changes in the model. An explanation of the causes is given by [Kelly et al., 1999]. Let us look at the Bayesian decision formula used in naive Bayes.

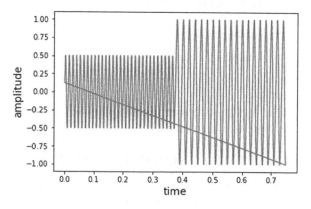

FIGURE 1.6   Concept drift in signals.

$$p(c/D) = \frac{P(D/c)\,p(c)}{p(D)}$$

posterior $\propto$ likelihood $\times$ prior

where D and c denote the data and class labels. A change can then be attributed to:

- Class prior $p(c)$ may change over time;
- Likelihood $p(D/c)$ might change;
- Posterior $p(c/D)$ can change.

The first two changes are called *virtual* concept drift whereas the last one is called a *real* concept drift. A further illustration of the three causes is given in Figure 1.7. We can see that the original decision boundary shifts its location due to changes in the class prior to (b) (distribution of the red class changes). In Figure 1.7(c), the decision

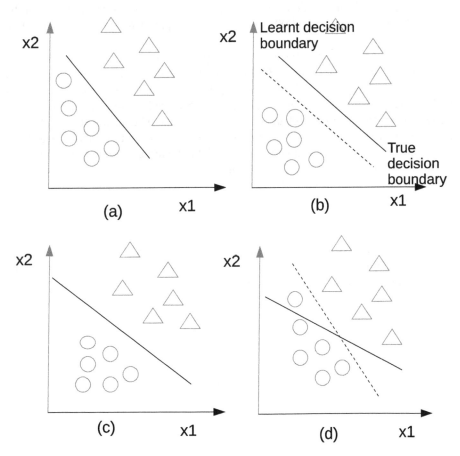

FIGURE 1.7 Effect on (a) original decision boundary due to change in (b) prior class, (c) likelihood, and (d) posterior class (Wang et al., 2018).

boundary remains unaltered. In Figure 1.7(d), the decision boundary changes drastically due to the change in the posterior probability. As a result, the previously learned decision function becomes useless or not effective. That is why handling concept drift due to a change in posterior probability is more difficult to tackle.

1.5.2 HANDLING CONCEPT DRIFT

In the literature, there are many ways to combat concept drift. Mostly, they are characterized in the following ways (Figure 1.8) (Gama, 2010):

In the *data management* way of handling concept drift, the learning mechanism makes use of memory to keep the data. In a *full-memory* mechanism, sufficient statistics are maintained over the entire data set by different kinds of weighting, and a model is trained on that. In a *partial-memory* mechanism, data is stored in the window and a model is built over it. In *detection methods*, the techniques focus on the learning mechanism such as how to detect the change and rate of change. These methods usually detect changes by monitoring them in the performance metrics or maintaining the time window. An example of the first approach is given by Klinkenberg and Joachims (2000), and the second approach is given in Kifer et al. (2004).

In the *adaptation method*, the learner is adapted to track changes in the data evolution either in an *informed way* or *blindly*. In the former case, the learner updates itself only when a change has indeed occurred, whereas in the latter case, the learner updates itself no matter whether a change has occurred or not. In *decision model management*, the technique characterizes the number of decision models kept in memory. It assumes that the data comes from several distributions and tries to learn a new model whenever a change is detected. The dynamic weighted majority algorithm (Kolter and Maloof, 2007) is an example of decision model management.

In the next sections, we will look at several examples of the change detection algorithms.

1.5.2.1 CUSUM Algorithm

Let us look at a classic change detection algorithm called CUSUM (CUmulative SUM) (Basseville and Nikiforov, 1993). This monitors the cumulative sum of the log-likelihood ratio for detecting a change. Consider a sequence of independent

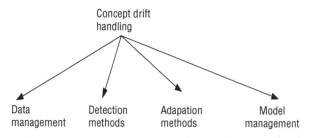

FIGURE 1.8 Methods of handling concept drift.

random variables x_t with pdf $p\theta(x)$. Our parameter is θ which before a change is θ_0 and after a change is θ_1. Assume θ_0 is known. The log-likelihood ratio is defined by

$$S_t = \log \frac{p_{\theta_1}(x)}{p_{\theta_0}(x)} \tag{1.3}$$

Formally, let S_t be the current cumsum of the log-likelihood and m_t the current minimum value of S_t. The CUSUM compares this difference with a threshold δ as follows:

$$g_t = S_t - m_t \geq \delta$$

$$S_t = \sum_{i=0}^{t} s_i$$

where

$$s_i = \log \frac{p_{\theta_1}(x)}{p_{\theta_0}(x)}$$

$$m_t = \min_{0 \leq j \leq t} S_j$$

So the change point is:

$$t_a = \min\left\{t : S_t \geq m_t + \delta\right\} \tag{1.4}$$

Note that the detection rule mentioned above is a comparison between the cumulative sum S_t and an adaptive threshold $m_t + \delta$. Due to the online update of m_t, the threshold is not only modified online, but also keeps a complete memory of the entire information contained in the historical observations. An example of applying CUSUM to detect changes in the cosine signal is shown in Figure 1.9. As we can see, at every negative and positive change, CUSUM demarcates the changes. However, since the change is slow, the actual alarm triggers after a delay (marked with a red dot). The bottom figure shows the cumulative changes in the positive and negative direction with color-coding. Note that the figures generated use a variant of the CUSUM presented above. They have been plotted using the cumulative sum of the difference of the values with a correction factor and then comparing with the threshold.

1.5.2.2 The Higia Algorithm

Higia (Garcia et al., 2019) is an online clustering algorithm for concept drift and novelty detection. The main idea of the algorithm is based on the clustering of evolving data streams using micro-clusters proposed in Aggarwal et al. (2003). Let us first understand the intuition behind the algorithm, then we will present its formal description. The Higia algorithm proceeds in two steps: an offline phase and an online phase. In the offline phase, it creates micro-clusters using an initial set of labeled data. Then, for each incoming data point, it calculates its distance from the centroid of the micro-clusters. If the distance is more than a user-specified threshold, it puts it in a buffer and, after enough points, declares it to be a novelty. The concept drift is declared when a new incoming data point increases the radius of the cluster.

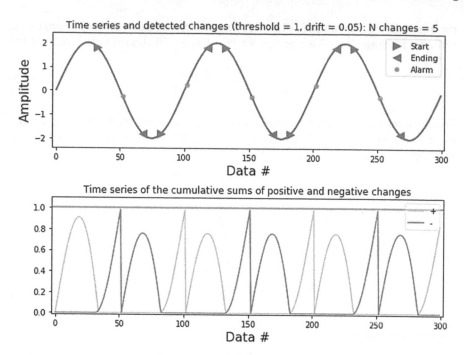

FIGURE 1.9　(Top) Time series data from cosine function along with marked beginning and end of changes. (Bottom) Cumulative sum plot of positive changes.

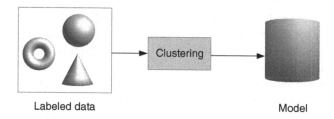

FIGURE 1.10　Higia Offline step.

Now we will look at the algorithm more formally. As shown in Figure 1.10, a set of initial labeled data is clustered following the same strategy as in Aggarwal et al. (2003) using the CluStream algorithm. The next step is an online phase as shown in Figure 1.11. In the online phase, the new incoming data point is fit to the model. Two things can happen here. If the new data point fits the model, two more situations arise. Either the new point is within the radius of a micro-cluster or outside. If the point is within the nearest micro-cluster, the data point is labeled with that of the micro-cluster. However, if it is outside but still within some factor (user-specified parameter) of the radius, the radius of the micro-cluster is updated and a new data point is labeled with that of this micro-cluster. The second case here is an example of concept drift. On the other side, if the data point does not fit the model well, it is put

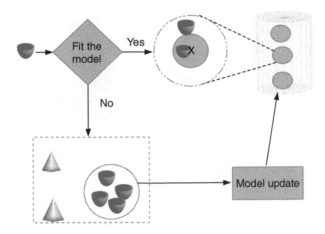

FIGURE 1.11   Higia Online.

into a buffer. When the buffer has received enough data points, the CluStream algorithm is run and the new cluster is put into the trained model.

The Higia is given in Algorithm 1.1. We will illustrate the algorithm more formally. Input to the algorithm is X_{tr} which is the new test point. T and k are user parameters. In the second step, ψ_k is a list of k nearest micro-clusters. If the majority of the clusters have the same label, we look for the centroid c_j of the nearest micro-cluster C_j and denote its radius by $radius(C_j)$. If the Euclidean distance of the point from the centroid is within the radius, update the cluster with the new point and classify the new point with the label of this cluster (lines 7–10). Otherwise, if the Euclidean distance is larger than T times that of the radius of the cluster, then create an extension of the cluster and classify the new data point with the same label (lines 11–13). If the above two scenarios have failed, the new data point in the buffer needs to classify the new data point as an *unknown* class level (lines 15–16).

Below, we show some empirical results of running the Higia algorithm on some benchmark data sets (Figure 1.12). Accuracy versus time is plotted in Figures 1.13

Algorithm 1. Higia: Online Phase

1: **input:** X_{tr}, T, k
2: Let ψ_k be a list of the k nearest micro-clusters to X_{tr}
3: **if** majority of ψ_k have the same label **then**
4:     Let C_j be the nearest micro-cluster to X_{tr}
5:     Let c_j be the centroid of C_j
6:     Let $radius(C_j)$ be the radius of C_j
7:     $dist \leftarrow EuclidianDistance(X_{tr}, C_j)$
8:     **if** $dist \leq radius(C_j)$ **then**
9:         update C_j with X_{tr}
10:         classify X_{tr} with the same label of C_j
11:     **else if** $dist \leq (radius(C_j) \times T)$ **then**
12:         create extension of C_j with centroid X_{tr} and radius 0.5
13:         classify X_{tr} with the same label of C_j
14: **else**
15:     add X_{tr} to buffer
16:     classify X_{tr} as unknown

FIGURE 1.12   Higia algorithm.

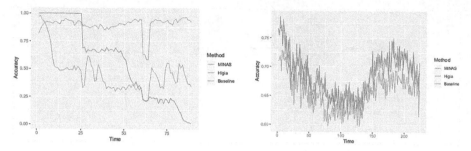

FIGURE 1.13 Accuracy over time plot for Higia and baselines. Image source: Garcia et al. (2019).

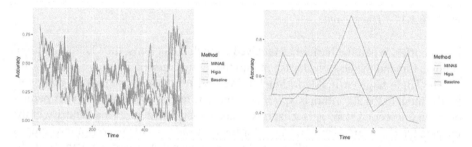

FIGURE 1.14 Accuracy over time plot for Higia and baselines. Image source: Garcia et al. (2019).

and 1.14. MINAS and kNN serve as a baseline algorithm. The performance of the Higia algorithm is shown in green. As we can see, Higia outperforms other algorithms over time in three out of four cases. kNN performs the worst because it does not learn over time.

1.5.2.3 Dynamic Weighted Majority Algorithm

Dynamic weighted majority (Kolter and Maloof, 2007) abbreviated as DWM is an algorithm for concept-drift tracking. It is based on the majority voting of experts. The fundamental idea behind DWM is that it dynamically creates and removes weighted experts based on their performance. The algorithm works according to four ideas: a train ensemble of learners; weighting them based on their performance; removing experts based on the performance; and adding new learners based on the global performance.

The DWM algorithm is presented in Algorithm 1.3. I will explain each step of the algorithm. There are m experts. The training pair is denoted by $\{x,y\}_n^1$ and there are c classes. The $\{e,w\}$ denotes the expert and its weight. In lines 1–3, we initialize the number of experts, create a new expert, and set its weight to 1. Then in the outermost for loop, we iterate over the stream one by one. First, we set the vector σ which denotes the sum of weighted predictions to 0 (line 5). In the second for loop, we iterate over the experts, making a prediction for each data point and decreasing its weight if it makes a mistake (lines 7–10). In line 11, we update the weighted prediction vector. Line 13 sets the global prediction of η_g weight. In the if part (lines 14–22), we normalize the weights. The reason for doing this is that some experts might be always right and their weights can go really high (due to the addition of weights in line 11). If the weight of an expert falls below the threshold α, it is removed from the set of experts (line 16). If the global prediction weight differs from the true label, that means the whole set of an expert is not good and this is the time to create a new expert. Finally, in lines 23–25, we train the new experts and return the global prediction weight in line 26. DWM is a general algorithm for coping with concept drift. One can use any online learning algorithm as the base learner, such as a decision tree, naive Bayes, or a neural network, within the framework.

Algorithm 1.3: Dynamic Weighted Majority

---

   input $\{x, y\}_n^1$: training data, $c \in \mathbb{N}, C \geq 2$: number of classes, α:threshold for deleting
:
       expert, β:factor for decreasing weights, q: time interval for removing, adding and
       weight update, $\{e, w\}_m^1$: set of experts and their weights, η_g, η_l: global and local
       predictions, $\sigma \in \mathbb{R}^c$: sum of weighted predictions for each class

1   $m \leftarrow 1$
2   $e_m \leftarrow Create\_New\_Expert()$
3   $w_m \leftarrow 1$
4   **for** $i \in \{1, \ldots, n\}$ **do**
5        $\sigma \leftarrow 0$
6   **for** $j \in \{1, \ldots, m\}$ **do**
7             $\eta_l \leftarrow Classify(e_j, x_i)$
8   **if** $\eta_l \neq y_i$ **and** $i \bmod q == 0$ **then**
9                $w_j \leftarrow \beta w_j$
10   **end**
11            $\sigma_{\eta_l} \leftarrow \sigma_{\eta_l} + w_j$
12   **end**
13        $\eta_g \leftarrow argmax_j \ \sigma_j$
14   **if** $i \bmod q == 0$ **then**
15            $w \leftarrow Normalize\_Weights(w)$
16            $\{e, w\} \leftarrow Remove - Expert(\{e, w\}, \alpha)$
17   **if** $\eta_l \neq y_i$ **then**
18                $m \leftarrow m + 1$
19                $e_m \leftarrow Create\_New\_Expert()$
20                $w_m \leftarrow 1$
21   **end**
22   **end**
23   **for** $j \in \{1, \ldots, m\}$ **do**
24            $e_j \leftarrow Train(e_j, x_i, y_i)$
25   **end**
26   *reutrn* η_g
27 **end**

1.6 DISCUSSION

Data stream mining is a challenging task because of the complex dynamics involved. In this chapter, I have presented some algorithms for some basic problems in streaming data such as filtering and counting. We also discussed how to sample from a data stream efficiently as well as handling concept drift. We can see that sampling and concept drift are some of the fundamental problems affecting other stream mining tasks such as classification, clustering, and novelty detection. Therefore, having a good sampler for streaming data off-the-shelf makes downstream tasks easier without worrying about devising a new algorithm for handling them separately. On the other hand, concept drift is another problem which often comes with class imbalance or novelty detection. DWM and CUSUM are classic algorithms that can be used to track concept drifts efficiently. Higia is a state-of-the-art model for such problems. There are still unsolved problems in big streaming data such as privacy preservation, handling incomplete and delayed information, and analysis of complex data to name a few (Krempl et al., 2014).

REFERENCES

Aggarwal, C. C. (2007). *Data Streams: Models and Algorithms*, volume 31. Springer Science & Business Media, Boston, MA.

Aggarwal, C. C., Watson, T. J., Ctr, R., Han, J., Wang, J., and Yu, P. S. (2003). A framework for clustering evolving data streams. In *VLDB*, Riva del Garda, Italy, pp. 81–92.

Basseville, M. and Nikiforov, I. V. (1993). *Detection of Abrupt Changes: Theory and Application*, volume 104. Prentice Hall, Englewood Cliffs, NJ.

Bhat, A. (2016). Use the bloom filter, luke! https://www.semantics3.com/blog/use-the-bloom-filter-luke-b59fd0839fc4/. Accessed September 30, 2019.

Chaudhuri, S., Ding, B., and Kandula, S. (2017). Approximate query processing: No silver bullet. In *Proceedings of the 2017 ACM International Conference on Management of Data*, pp. 511–519. ACM, New York.

Cormode, G. and Muthukrishnan, S. (2004). Improved data stream summaries: The count-min sketch and its applications. *Journal of Algorithms*, 55(1):58–75.

Das, H., Dey, N., and Balas, V. E. (2019). *Real-Time Data Analytics for Large Scale Sensor Data*. Academic Press, London, UK.

Dey, N., Das, H., Naik, B., and Behera, H. S. (2019). *Big Data Analytics for Intelligent Healthcare Management*. Academic Press, Boston, MA.

Flajolet, P., Fusy, É., Gandouet, O., and Meunier, F. (2007). Hyperloglog: The analysis of a near-optimal cardinality estimation algorithm. In *Discrete Mathematics and Theoretical Computer Science*, pp. 137–156. DMTS, Nancy.

Flajolet, P. and Martin, G. N. (1985). Probabilistic counting algorithms for data base applications. *Journal of Computer and System Sciences*, 31(2):182–209.

Gama, J. (2010). *Knowledge Discovery from Data Streams*. Chapman and Hall/CRC, Boca Raton, FL.

Garcia, K. D., Poel, M., Kok, J. N., and de Carvalho, A. C. (2019). Online clustering for novelty detection and concept drift in data streams. In *EPIA Conference on Artificial Intelligence*, pp. 448–459. Springer, Cham, Switzerland.

Kelly, M. G., Hand, D. J., and Adams, N. M. (1999). The impact of changing populations on classifier performance. In *Proceedings of the Fifth ACM SIGKDD International Conference on Knowledge Discovery and Data Mining*, pp. 367–371. ACM, New York.

Kifer, D., Ben-David, S., and Gehrke, J. (2004). Detecting change in data streams. In *Proceedings of the Thirtieth International Conference on Very Large Data Bases (VLDB '04)*, volume 30, pp. 180–191. VLDB Endowment.

Klinkenberg, R. and Joachims, T. (2000). Detecting concept drift with support vector machines. In *Proceedings of the Seventeenth International Conference on Machine Learning (ICML)*, pp. 487–494. Morgan Kaufmann, San Francisco, CA.

Kolter, J. Z., and Maloof, M. A. (2007). Dynamic weighted majority: An ensemble method for drifting concepts. *Journal of Machine Learning Research*, 8:2755–2790.

Krempl, G., Žliobaite, I., Brzeziński, D., Hüllermeier, E., Last, M., Lemaire, V., Noack, T., Shaker, A., Sievi, S., Spiliopoulou, M. et al. (2014). Open challenges for data stream mining research. *ACM SIGKDD Explorations Newsletter*, 16(1):1–10.

Li, K. and Li, G. (2018). Approximate query processing: What is new and where to go? *Data Science and Engineering*, 3(4):379–397.

Rajaraman, A. and Ullman, J. D. (2011). *Mining of Massive Datasets*. Cambridge University Press, Cambridge, UK.

Wang, S., Minku, L. L., and Yao, X. (2018). A systematic study of online class imbalance learning with concept drift. *IEEE Transactions on Neural Networks and Learning Systems*, 29(10):4802–4821.

2 Decoding Common Machine Learning Methods

Agricultural Application Case Studies Using Open Source Software

Srinivasagan N. Subhashree, S. Sunoj,
Oveis Hassanijalilian, and C. Igathinathane

CONTENTS

2.1 INTRODUCTION

Machine learning (ML) applications are ubiquitous in the present digital age. The ML methods are gaining attention due to their ability to solve complex problems and provide a decision support system from large volumes of data. Software packages are available for different ML methods in the form of modules in various programming languages (e.g., Python, R, MATLAB). Sufficient online resources and publications are available discussing the theoretical framework of ML methods and practical usage of the software modules in various platforms. However, to our knowledge, resources dealing with the concepts and mechanisms of the ML methods and the sequence of intermediate operations performed inside the ML algorithms with a working code are not available. The ML methods are usually presented as "black-box" products providing no insight into the mechanism or allowing for modification to suit user needs. It will always be desirable to develop ML algorithms from funda-mental principles, especially using open source free software, to fit the processing and analyzing requirements of the user. Therefore, it is proposed to focus on decod-ing the two common ML methods and develop working codes using open source R software and demonstrate the applications.

Among the various fields that successfully employ ML methods, modern agri-culture is one of the sectors that produces an enormous amount of data from dif-ferent equipment components, such as yield monitors, seed and fertilizer equipment, soil moisture sensors, cameras, and global positioning systems (GPS). Another data domain is a variety of images from satellites, unmanned aerial vehi-cles, spectral camera (e.g., hyper, IR, NIR, NDVI) imagery, LiDAR point clouds, field mounted PhenoCams, as well as simple handheld color digital cameras. However, as the source of these data varies greatly, their type and quality will also vary accordingly. Furthermore, since agriculture deals with biological systems, there exists considerable variability, which presents a challenging scenario to be tackled through common/simple data analysis procedures. The ML methods are generally applied in analyzing such diverse data to make informed decisions in agriculture. Therefore, in this chapter we used two agricultural field application datasets, such as soybean aphid identification and weed species classification, for the demonstration of decoded ML methods.

Of several agricultural applications, the motivation for choosing these two datasets is due to the complexity of the problems. In the soybean aphid dataset, the other objects (exoskeletons and leafspots) exhibit features similar to aphids; hence classifying aphids is complicated and requires the selection of suitable features (individual or a combination). In the weed dataset, the classification of weeds into grassy and clover-shaped species at the "just emerged" stage is a difficult and tedious task, as both species display almost similar shape (two to four emerged leaves). The two datasets used in this chapter were primarily in the form of images, which were processed using ImageJ (Rasband, 2019; Schindelin et al., 2012), an open source and free image processing software, to extract features suitable for ML application.

Even though employing ML classification using available packages is a simple and plug-and-play methodology, it is vital to realize that it is a blindfold approach. The user supplies the input data to the model and receives output while being oblivious of the process behind the packaged ML methods. The methods in such packages, commonly referred to as "black-boxes," are built with the highest efficiency specific to their ecosystem, and when required it is not always possible to customize them to a different ecosystem. It is always desirable to decode and reconstruct the ML methods to overcome these limitations.

The primary advantage of such user-coded models is that the user is fully aware of the mechanism, the mathematical principles behind the algorithm, and the models' limitations. The models when reconstructed in open source software will be cost-effective and have less constraint and complexity. Furthermore, user-coded models can be customized based on the data, could potentially improve the model performance more than the black-box approaches, and provide knowledge and training to the users (e.g., students and researchers) as it is developed from scratch.

Given this background and requirement to develop decoded ML algorithms, the objectives of this chapter were to: (i) decode two of the common ML classification methods using open source software ImageJ (image processing and data extraction) and R (data handling and model development); (ii) employ the decoded algorithm for demonstrating classification using field data of agricultural applications, such as soybean aphids identification and weed species classification; and (iii) compare the performances of the decoded algorithm with the existing black-box approaches.

2.2 LITERATURE REVIEW

The superior power of ML algorithms is the ability to solve problems that were not possible to address by simpler mathematical approaches with explicit programming. The overall process of ML has the following three steps: (i) finding patterns from input data, (ii) building an ML model, and (iii) predicting the response using the ML model (Das et al., 2020). Recent applications of ML algorithms can be found in cutting-edge technologies from Netflix and Amazon recommendations (Gomez-Uribe and Hunt, 2016), fraud detection for financial institutions (Kültür and Çağlayan, 2017), driverless cars, and autonomous vehicles (Choi and Cha, 2019; Ali Alheeti and McDonald-Maier, 2018) to biomedical and healthcare fields (Das et al., 2020b;

Dey et al., 2019), including agriculture (Liakos et al., 2018; Zhou et al., 2019). The common theme among these examples is the availability of rich datasets.

Some of the successful agricultural applications of ML include crop yield prediction (Löw et al., 2018), inputs quantification (Tantalaki et al., 2019), phenology and crop health monitoring (Behmann et al., 2015), weeds classification and geo-location (dos Santos Ferreira et al., 2017), animal behavior and health monitoring (Borchers et al., 2017), and plant disease detection and classification (Ferentinos, 2018).

Most of these applications used software packages that are available for different ML methods in the form of black-box modules in various programming languages, such as Scikit-learn and TensorFlow in Python (Pedregosa et al., 2011; Abadi et al., 2015; Sarkar et al., 2018); nnet, and caret in R (Ramasubramanian and Singh, 2017; Kuhn, 2019); and Statistic and ML toolbox in MATLAB (Martinez and Martinez, 2015). All of these packages are great for training, developing, and testing ML models, but they come with their pros and cons. Python and R (R Core Team, 2019) are preferred among industry, since they are free, open source, and work cross-platform; while MATLAB, a commercial software that, as well as requiring an annual subscription, is commonly used in academia.

Among the agricultural applications of ML, pest identification and weed species classification are a few common issues being addressed. For these agricultural applications, the available ML packages developed for addressing generic applications are being adapted and tested, but a reliable solution is still lacking. This is possibly due to the lack of resources that facilitate agricultural researchers in understanding the mechanism of ML methods. Better clarity on this aspect will help in customizing the methods to suit a specific agricultural application. Therefore, this chapter focuses on handling agricultural field data examples and provides an overview of the application of user-developed, decoded, commonly used ML methods. Particularly in the agricultural domain, this chapter could possibly change the perception of the straightforward "application" mindset among several researchers to a "development" mindset. Such an impact will help in developing solutions for other specific agricultural problems, not limited to the aforementioned common applications.

2.3 MATERIALS AND METHODS

2.3.1 OVERALL ML MODEL DEVELOPMENT PROCESS

The overall processing stages for the decoded ML model development in this chapter are outlined (Figure 2.1) and subsequently described. The data required for this process were generated from the digital images acquired from the two different agricultural applications of identification and classification (soybean aphids and weed species). The actual input data for the ML models need to be in a structured format (with the same data type across features). We used ImageJ (Version 1.52r) for processing and extracting data (Rasband, 2019; Schindelin et al., 2012). ImageJ is an open source, java-based, and free image processing software that allows for the development of task-specific and user-coded applications in the form of plugins (Igathinathane et al., 2009). ImageJ also supports plugin development in several other languages (e.g., JavaScript, Python, Ruby, Clojure, Scala). The raw color images were preprocessed using ImageJ's thresholding methods to obtain binary

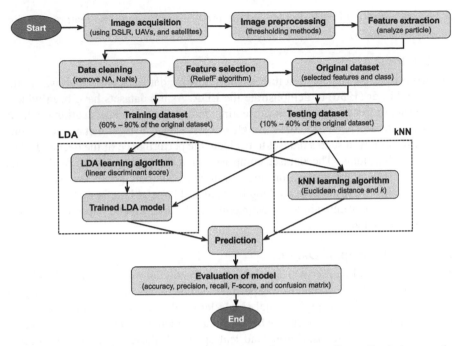

FIGURE 2.1   Overall process flow diagram of decoding common linear discriminant analysis (LDA) and k-nearest neighbor (kNN) machine learning methods.

images (suitable for further processing) for feature extraction. ImageJ offers a several measurements (features) option from the binary image, such as the object area, major and minor axes length, width and height of bounding box, and Feret dimensions, to name a few. These basic measurements can be used to derive additional shape features to address the requirement of different applications (Igathinathane et al., 2008; Sunoj et al., 2017, 2018; Du and Sun, 2004).

The evaluated shape features were processed using R, an open source and free statistical software (R Core Team, 2019). A user-coded R program was developed to process the data and build the ML models based on linear discriminant analysis (LDA) and k-nearest neighbor (kNN) algorithms (Figure 2.1). The initial data processing included data cleaning, feature selection, and data splitting into training and testing datasets. The ML models were decoded showing various intermediate processes that run inside the ML algorithms. The results of the user-coded models were compared with the black-box models, from readily available R packages, based on the model performance and the time taken for processing.

2.3.2 DATA COLLECTION

In this chapter, we use the following three datasets: (i) iris, (ii) soybean aphids, and (iii) weed species. The iris dataset was used for the demonstration and decoding of different ML classification methods, while the other two datasets were used for showing the application of ML classification methods on agricultural field applications.

A brief description of each dataset and the process of obtaining the features are discussed subsequently.

2.3.2.1 Iris Dataset

The iris is a multivariate dataset that is well-known in data analysis which was introduced by Fisher (1936) to demonstrate the LDA. As iris datasets have been widely used, even in recent literature, for presenting and testing new classification methods (Kamarulzalis and Abdullah, 2019; Singh and Srivastava, 2019; Bai et al., 2019), this study also employed such a dataset to evaluate the performance of the decoded LDA and kNN algorithms. The dataset comprises of 150 observations from three iris flower species (setosa, versicolor, and virginica). Each data point consists of four features (units cm), such as sepal length, sepal width, petal length, and petal width. The dataset is preinstalled in R software command, which can be imported using the data(iris) command.

2.3.2.2 Soybean Aphid Dataset

The soybean aphid dataset was obtained from the digital images of aphid infected soybean leaves from a greenhouse experiment (Sunoj et al., 2017). The color digital images were acquired using a digital single-lens reflex camera (Canon EOS Rebel T2i). Identifying aphids on the soybean leaf is challenging due to the presence of other objects such as exoskeletons and leaf spots. All these objects exhibit only a subtle difference in color, but a clear difference in shape (Figure 2.2A). Therefore, the images were preprocessed to extract shape features of individual objects using ImageJ. The sequence of image preprocessing steps involved in shape-feature extraction is presented in Figure 2.2.

The acquired color digital image (Figure 2.2A) was denoised by applying a median filter of two-pixel radius (Figure 2.2B). The filtered image was converted into a grayscale image to facilitate the thresholding process (Figure 2.2C). ImageJ offers 17 optimized thresholding schemes out of which a Yen thresholding scheme (Yen et al., 1995) was suitable as it appropriately selected the necessary portions (Figure 2.2D). Applying the thresholded scheme produced the binary image (Figure 2.2E). The artifacts and rough object boundary in the binary image were removed by applying the "Open" morphological operation (Figure 2.2F). It is essential to segment the overlapping objects into individual objects before feature extraction. The watershed segmentation offered by ImageJ resulted in oversegmentation; therefore, the overlapping objects were manually segmented by drawing a one-pixel-thick white line using ImageJ's "paintbrush tool" (Figure 2.2G). The segmented objects were subjected to ImageJ's "Analyze Particles" to derive various shape features (see Section 2.3.3) used in the analysis.

2.3.2.3 Weed Species Dataset

The weed species dataset was obtained from digital images acquired in a greenhouse for identifying the potential weed species grown in the fields. The weeds were grown in soil samples that naturally had seeds, which were transferred to the greenhouse in plastic trays from organic field experiment plots, and imaged right after emergence. Visually counting and categorizing the weed species at an early stage from multiple

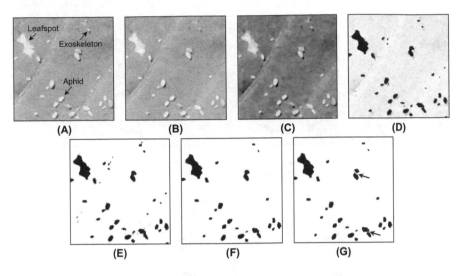

FIGURE 2.2 Image preprocessing steps for feature extraction from the digital images of soybean leaves: (A) Color digital image of the soybean leaf (zoomed-in) showing aphids, exoskeletons, and leaf spots; (B) Median filtered (radius = 2 pixel) image to remove noise; (C) Grayscale image; (D) Thresholded regions in dark gray or red obtained by applying ImageJ's Yen thresholding scheme; (E) Binary image obtained after thresholding; (F) Artifacts removed from image after applying single "Open" operation; and (G) Manually segmented image by drawing one-pixel-thick line using ImageJ's paintbrush tool.

trays is challenging. Therefore, an image-based approach could be developed from the shape features derived from the weed species images. A comprehensive review of weed identification through image processing techniques also showed a lack of literature on identifying the "just emerged" weed species from digital images (Wang et al., 2019).

The digital images were captured using a consumer-grade color digital camera (Samsung NX3300). A user-developed JavaScript plugin was employed to process the images for shape features extraction (Figure 2.3). The input color digital image consisted of two weed species that clearly exhibited differences in shape (Figure 2.3A). The color digital image was split into Lab channels, out of which the "a" channel displayed a clear difference in gray intensities between the background soil and foreground plants (Figure 2.3B). Applying the ImageJ's Yen thresholding scheme on the "a" channel selected the foreground plant portions (dark gray or red) of the image (Figure 2.3C). The thresholded binary image captured even tiny plant portions, which were not distinctly visible in the input image (Figure 2.3D).

The weeds were grown in a random arrangement that sometimes resulted in an overlap between objects. In addition, since the plants were at an early growth stage, the thresholding operation disconnected a few objects. To overcome these issues, the binary image was visually inspected and the overlapping objects were manually segmented by drawing a one-pixel-thick white line (green arrows, Figure 2.3E). Similarly, to bridge the disconnected objects, a one-pixel-thick black line was manually drawn at appropriate locations (dotted-arrows, Figure 2.3E).

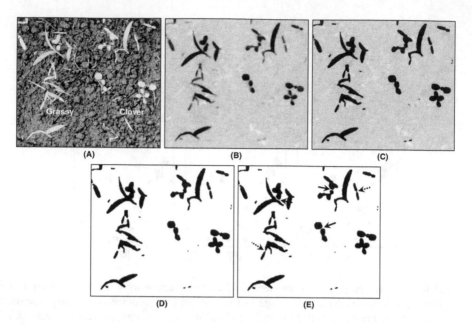

FIGURE 2.3 Image preprocessing steps for feature extraction from an image of weed species seedlings. (A) Input color digital image; (B) The "a" color channel in Lab color space; (C) Thresholded regions in dark gray or red obtained through ImageJ's Yen thresholding scheme; (D) Binary image after thresholding; and (E) Final preprocessed binary image (after correction) used for shape feature extraction.

Carefully delineating the individual object before measuring shape features is essential because the performance of ML algorithms solely depends on the quality of the input data. The shape features discussed subsequently (see Section 2.3.3) were extracted from the final preprocessed image (Figure 2.3E; without colored arrows) and used as the training data for ML models.

2.3.3 SHAPE FEATURES EXTRACTION

Shape features are dimensionless parameters, which are not affected by image resolution, and were used as measures in classifying objects. ImageJ offers four shape features, such as aspect ratio, circularity, roundness, and solidity (Equations (2.1)–(2.4)), along with various other measurements. Those measurements were used for deriving other shape features reported in the literature (Igathinathane et al., 2008; Sunoj et al., 2017; Du and Sun, 2004) and are presented as follows (Equations (2.5)–(2.13)):

$$\text{Aspect ratio} = \frac{a}{b} \qquad (2.1)$$

$$\text{Circularity} = \frac{4\pi A}{P^2} \qquad (2.2)$$

$$\text{Roundness} = \frac{4A}{\pi a^2} \qquad (2.3)$$

$$\text{Solidity} = \frac{A}{A_C} \qquad (2.4)$$

$$\text{Reciprocal aspect ratio} = \frac{b}{a} \qquad (2.5)$$

$$\text{Rectangularity} = \frac{A}{H_{BB} \times W_{BB}} \qquad (2.6)$$

$$\text{Feret major axis ratio} = \frac{D_F}{a} \qquad (2.7)$$

$$\text{Area ratio} = \frac{A}{D_F \times D_{minF}} \qquad (2.8)$$

$$\text{Diameter range} = D_F - D_{minF} \qquad (2.9)$$

$$\text{Eccentricity} = \sqrt{1 - \frac{(b/2)^2}{(a/2)^2}} \qquad (2.10)$$

$$\text{Area-perimeter ratio} = \frac{A}{P^2} \qquad (2.11)$$

$$\text{Hollowness} = \frac{A_C - A}{A_C} \times 100 \qquad (2.12)$$

$$\text{Feret aspect ratio} = \frac{D_F}{D_{minF}} \qquad (2.13)$$

where a is the major axis length (pixel); b is the minor axis length (pixel); A is the area (pixel^2); P is the perimeter (pixel); A_C is the convex area (pixel^2); H_{BB} is the bounding box height (pixel); W_{BB} is the bounding box width (pixel); D_F is the Feret diameter, also called the maximum caliper diameter (pixel); and D_{minF} is the minimum Feret diameter, also called the minimum caliper diameter (pixel).

2.3.4 DATA CLEANING

Cleaning the dataset is essential since even robust ML methods underperform while processing the improper and uncleaned data. On average, about 60% of the time in ML is spent on cleaning and unifying the obtained data (Chu et al., 2016). Cleaning the data usually refers to filtering and editing the data to make it easier to explore, understand, and model. Data filtering is removing unnecessary, unwanted, and irrelevant data from the dataset, while editing refers to modifying the data into a desirable and accessible format. Data cleaning was performed in R using conventional techniques, functions, and commands.

The common steps followed include checking missing values, replacing or dropping missing values, and modifying the variable type. In our datasets, the presence of missing values were checked using the is.na() and is.nan() functions. There were no missing values in our dataset; however, if it exists, the user may choose to replace the missing data with a specified value or omit the entry. To replace a value with 0, the function data[is.na(data)] <- 0 can be used; while to omit the data, the function na.omit() can be used. Once the data is cleaned, it is desirable to convert the data type into an appropriate form (e.g., class as factor, feature values as numeric). These can be performed using R coercion functions, such as as.numeric(), as.integer(), as.character(), and as.factor().

2.3.5 FEATURE SELECTION

The datasets considered for ML modeling consists of many features (predictor variables) and an outcome (dependent variable). Mostly only a group of the features are relevant, i.e., it provides information to predict the outcome while the remaining are irrelevant features that are not informative. Developing ML models using all the available features results in computational burden and reduced accuracy. Therefore, the relevant features have to be carefully identified and selected to build the ML models while discarding others (Das et al., 2020a).

Feature selection is often confused with dimensionality reduction, as in those conducted using methods such as principal component analysis and single value decomposition. Both feature selection and dimensionality reduction methods seek to reduce the number of features or attributes in the dataset. However, the dimensionality reduction achieves this by creating new combinations of all the features (Das et al., 2020b), while the feature selection mutes the irrelevant features which are less useful in predicting the outcome variable. In general, fewer relevant features are desirable since it produces simpler models and provides a cost-effective and computationally efficient solution. Therefore, feature selection methods are more beneficial and useful in developing ML models than the dimensionality reduction methods. Feature selection methods can be classified into three general methods, such as filter, wrapper, and embedded methods which are briefly described.

2.3.5.1 Filter Methods

The filter method of feature selection assigns a score to each feature in the dataset by applying statistical information and correlation measure. The features are ranked based on the score, which implies the order of importance to the output variable. The features are either selected or discarded based on the rank. Some examples of filter selection methods are information gain (Joachims, 1998), chi-square test (Jin et al., 2006), and Relief (Kira and Rendell, 1992).

2.3.5.2 Wrapper Methods

The wrapper method uses an inbuilt ML classification and regression algorithm, where the combination feature subsets are created based on approaches such as forward or backward selection and heuristic selection. These selected features are given to the specific predictive model algorithm, and the scores are assigned to the feature subset based on the accuracy of the model. Similarly, each feature subset is analyzed, and the one yielding the highest model accuracy is selected. An example of a wrapper method is the forward and backward selection algorithm (Mao, 2004).

2.3.5.3 Embedded Methods

The embedded method of feature selection learns the features that contribute the most to the accuracy of the model. These methods are also known as the regularization or penalization methods, where the coefficients of the irrelevant features are penalized and removed from the model. Some of the examples of the regularization algorithms are Lasso (Li et al., 2006), elastic net (Zou and Hastie, 2005), and ridge regression (Endelman, 2011).

Among these feature selection methods, the Relief algorithm, which falls under filter methods, is mathematically straightforward, provides a measure of the influence for the features, and can deal with multi-classes in the outcome variable. Moreover, it is the only feature selection method that is capable of detecting the feature dependencies using the simple nearest neighbor approach (Urbanowicz et al., 2018). Therefore, the Relief algorithm was chosen for feature selection in the two considered agricultural applications.

2.3.5.4 Relief Algorithms

The Relief algorithm is a family of three algorithms, such as basic Relief, ReliefF, and RReliefF. The basic Relief algorithm can only handle the classification of two classes; the ReliefF, which is an extension of basic Relief, can handle multi-classes; and RReliefF can handle continuous regression classes. All three algorithms have the same underlying working mechanism. Since this chapter deals with classification with ML for a dataset containing multi-classes that are categorical, only the ReliefF method was appropriate. However, to follow the algorithm working principle, a manual example using the basic Relief algorithm for binary class data is demonstrated in this section since the manual demonstration using the ReliefF for multi-classes is tedious. Studies detailing the ReliefF algorithm can be found elsewhere (Kononenko, 1994; Robnik-Šikonja and Kononenko, 2003).

The pseudocode (Algorithm 2.1) decodes the basic Relief algorithm. Weights are initialized to zero for all the features in the data. The Relief algorithm identifies the two nearest neighbor, nearest hit (H) and miss (M), using Manhattan distance (rectilinear distance as sum of projections as opposed to the Euclidean distance as the direct straight line distance between points) calculated between the randomly selected instance (R_i) and the other instances in the dataset. The H instance refers to the instance that belongs to the same class and at the least distance to the random instance. The M refers to the instance that belongs to a different class and at the nearest distance to the random instance. The weight revisions are divided by the number of iterations (z) to normalize the weights between −1 and +1. As a rule of thumb, z is calculated as the square root value of the total number of observations in the dataset (n).

Algorithm 2.1 The pseudocode of the basic Relief algorithm

1: **procedure** BASIC_RELIEF(dataset)
2:    $W[F_j] = 0$ ▷ Set all weights of the features to zero
3:    $z = \sqrt{n}$ ▷ n - number of observations in the dataset
4:    **for** (i to z) **do**   ▷ z - number of iteration for weights revision
5:       select a random instance (R_i)
6:       find nearest hit (H) and miss (M) instances   ▷ using Manhattan distance
7:       **for** (1 to j) **do**   ▷ j is the number of features; $W[F_j]$ revised
8:          $W[F_j] = W[F_j] - \text{diff}(F_j, R_i, H)/z + \text{diff}(F_j, R_i, M)/z$
9:       **end for**
10:   **end for**
11: **end procedure**
while,

$$\text{diff}(F, R_i, H) = \frac{|\text{value}(F_j, R_i) - \text{value}(F_j, H)|}{\max(F_j) - \min(F_j)}$$

$$\text{diff}(F, R_i, M) = \frac{|\text{value}(F_j, R_i) - \text{value}(F_j, M)|}{\max(F_j) - \min(F_j)}$$

where, F_j are the different features in the data set; $F_j = F_1, F_2, \ldots, F_n$.

The idea of revising the weights of features is based on the dissimilarity of the random instance with H and M. A significant difference between the nearest hit and the random instance indicates that the features separate two instances belonging to the same class, which is not desirable; therefore, the weights are decreased. However, a significant difference between the nearest miss and the random instance implies that the features separate the two instances belonging to different classes, which is desirable; therefore, the weights are increased.

The following R program developed and based on the decoded Relief algorithm (Algorithm 2.1) estimates the weights of the features as outcomes (Figure 2.4). For comparison, the direct black-box approach was also used as an illustration. These weights are further used to determine the most significant features.

For demonstration (Figure 2.4), consider a sample dataset with 20 total observations ($n = 20$), three features (F_1, F_2, and F_3), two classes (classes 1 and 2), and five instances ($R_1, R_2, \ldots, R_5$). The relief algorithm is employed to this sample dataset to determine the most influential features, amongst F_1, F_2, and F_3, ranked according to the weights used in classifying the dataset into classes 1 and 2.

The features producing the highest weights are considered to be the most influential. In the case of sample data, the feature that ranked the highest, based on the weights, is $F_2(0.93)$, followed by $F_1(0.60)$ and $F_3(-0.50)$, respectively. The feature ranking obtained using the in-built (black-box) function attrEval(), under the R package CORElearn developed and based on the basic Relief algorithm, produced different weights (Figure 2.4).

2.3.6 DATA SPLITTING

The success of building computationally sound ML models with high prediction accuracy is based on its ability to generalize and be applied to different scenarios.

```
 1  # Load sample data ---------------------------------------------------------
 2  data <- read.table(header = T, text = '
 3                     F1  F2  F3  Class
 4            R1        8   2   2   1
 5            R2        7   2   5   1
 6            R3        3   8   2   2
 7            R4        6   9   5   2
 8            R5        4   9   9   2  ')
 9
10  ## Assigning and initializing weights to the features -----------------------
11  # Let the weights of features F1, F2, and F3, be wF1, wF2, and wF3
12  # Initializing weights to zero
13  wF1 = 0; wF2 = 0; wF3 = 0
14  n <- nrow(data) * ncol(data) # number of observations
15  z <- round(sqrt(n)) # number of iterations
16  for(i in 1:z){
17      require(dplyr)
18      randat <- sample_n(data, 1) # randomly selected instance
19      featdat <- data[1:3] # contains only features sans the class
20      man_dis <- array(0, c(nrow(data), 2)) # new array for Manhattan distances
21
22      # Calculating Manhattan distance ------------------------------------
23      # Manhattan distance between the features and randomly selected instances
24      for(j in 1:nrow(data)){
25          man_dis[j,1] <- (abs(randat[[1]] - featdat[j,1]) + abs(randat[[2]]
26                          - featdat[j,2]) + abs(randat[[3]]-featdat[j,3]))
27          man_dis[j,2] <- data$Class[j]
28          } # end for - Manhattan
29
30      # Removing the randomly instance from data --------------------------
31      man_dis1 <- man_dis[-which(man_dis[,1]==0),]
32      data1 <- data[-which(man_dis[,1]==0),]
33      data1$sum <- man_dis1[,1]
34
35      # Finding the nearest hit -------------------------------------------
36      nearHit <- subset(data1, data1$Class == randat$Class)
37      indH <- match(min(nearHit$sum), nearHit$sum)
38      nearHit <- nearHit[indH,]
39
40      # Finding the nearest miss ------------------------------------------
41      nearMis <- subset(data1, data1$Class != randat$Class)
42      indM <- match(min(nearMis$sum), nearMis$sum)
43      nearMis <- nearMis[indM,]
44
45      # Finding weights of the features -----------------------------------
46      wF1 = wF1 - ((abs(randat$F1 - nearHit$F1)/(max(data$F1) - min(data$F1)))/z) +
47                  ((abs(randat$F1 - nearMis$F1)/(max(data$F1) - min(data$F1)))/z)
48      wF2 = wF2 - ((abs(randat$F2 - nearHit$F2)/(max(data$F2) - min(data$F2)))/z) +
49                  ((abs(randat$F2 - nearMis$F2)/(max(data$F2) - min(data$F2)))/z)
50      wF3 = wF3 - ((abs(randat$F3 - nearHit$F3)/(max(data$F3) - min(data$F3)))/z) +
51                  ((abs(randat$F3 - nearMis$F3)/(max(data$F3) - min(data$F3)))/z)
52  } # end for - iterations
53
54  # Final weights after 'z' iterations ----------------------------------------
55  wAll <- as.data.frame(c(wF1, wF2, wF3))
56  rownames(wAll) <- c("F1", "F2", "F3")
57  colnames(wAll) <- "Weights"
58  > wAll
59          Weights
60  F1    0.6000000
61  F2    0.9285714
62  F3   -0.5000000
63
64  # Finding the most influential weights ---------------------------------------
65  srt <- sort(wAll$Weights, index.return = T, decreasing = T)
66  srt <- c(srt$ix[1], srt$ix[2], srt$ix[3]) # Sorted weight results
67  > rownames(wAll)[srt]
68  [1] "F2" "F1" "F3"
69
70  # Feature ranking using in-bulilt function -----------------------------------
71  require(CORElearn)
72  (estReliefF <- attrEval("Class", data, estimator="Relief", ReliefIterations=z))
73  > sort(estReliefF, decreasing = T)
74      F2    F1     F3       # Ranking of wAll is maintained
75    0.75  0.00  -0.75       # Actual weights might vary from wAll
```

FIGURE 2.4   Manually reconstructed basic Relief algorithm with a sample data.

Therefore, in ML, the available labeled datasets are partitioned into training and testing datasets. The training data subset is used to train the ML model by recognizing the pattern in the dataset and learning the relationship between the features and their respective class labels, while the testing data subset, which serves as a proxy to the incoming new dataset in the future, is used to estimate the generalization quality of the ML model (prediction accuracy) beyond the training dataset (Das et al., 2020c).

There are different data splitting methods, such as simple random sampling, trial-and-error, cross-validation, systematic sampling, convenience sampling, and stratified sampling. Among these methods, the simple random sampling is commonly used as it is easy and efficient to implement; however, the simple random sampling works more efficiently for datasets that had uniform distribution (Reitermanova, 2010). The Kolmogorov–Smirnov test conducted, using the R function ks.test(), revealed that most of the features selected from the datasets of soybean aphids, weed species, and iris were uniformly distributed ($p > 0.01$), hence making data subsetting valid for analyses. Therefore, in this study the simple random sampling method was used to select the data for training and testing from the original dataset.

Random sampling was performed using the R function createDataPartition() under the caret package. The split ratio of the dataset for training and testing usually is in the range of 60–90% and 40–10%, respectively. The split ratio is selected based on the prediction accuracy of the ML classifier model using the validation data, which is the subsample random selection of the training data. The iris dataset was used to demonstrate the working of the function createDataPartition() in splitting the original dataset into training and testing subset data. A ratio of 70% (training subset) and 30% (testing subset) was considered for data splitting. The R code snippet that achieves package loading, data partitioning, splitting into training and testing data are given below:

```
library(caret) # R package for Classification And REgression Training
# Data partitioning with 70% splitting
splitDat <- createDataPartition(y = data$Species, p = 0.7,
    list = FALSE)
trainDat <- data[splitDat,] # A split 70% assigned as training data
testDat <- data[-splitDat,] # The rest of 70% i.e., 30% assigned as
    testing data
```

2.3.7 THE ML METHODS

The two major categories of ML classification are unsupervised and supervised algorithms. Unsupervised ML performs classification on the unlabeled data (only input variables and no output variables with identified labels passed to the algorithm). The unsupervised algorithm learns the underlying structure and distribution in the data and clusters them as groups (e.g., clustering and association), while the supervised ML performs classification and regression tasks on the labeled data (input variables and labeled output variables passed to the algorithm), which helps the models to understand the relation between the input and output variables (Das et al., 2015, 2018).

Classification models are employed if the outcome variable is categorical, and regression models are used if continuous. Practical machine learning applications predominantly use supervised classification. The agricultural application datasets used in this chapter (soybean, aphids, and weed species) are labeled and have a categorical output variable. Therefore, common supervised classification ML algorithms, such as LDA and kNN, were considered, decoded, methods developed and based on their algorithms, and discussed.

2.3.7.1 Linear Discriminant Analysis

The LDA is a popular technique for dimensionality reduction and supervised classification and was originally developed to address the two-class problem (Fisher, 1936). It provides two regions separated by a line which aids in the classification of data and the regions, and the separation line is defined by the linear discriminant score function. This method was later generalized for multiclass problems (Rao, 1948). A linear discriminant score function is employed by the LDA method to predict the class labels upon feeding the test data instance. The discriminant score function is derived from Bayes theorem, which determines the class labels based on the probability of each class and the probability of the test data instance belonging to the class (Equation (2.14)).

$$P\left(\pi_i | \vec{x}\right) = \frac{P\left(\pi_i, \vec{x}\right)}{P\left(\vec{x}\right)} \tag{2.14}$$

where $P\left(\pi_i | \vec{x}\right)$ is the probability of test data instance $\vec{x}$ belonging to π_i class, $P\left(\pi_i, \vec{x}\right)$ is the joint probability of π_i and $\vec{x}$, and $P\left(\vec{x}\right)$ is the prior probability of $\vec{x}$.

The linear discriminant score function of Bayes theorem is expressed in the form of $y = ax + b$, where a and b are the parameters obtained from the training data and their respective classes, while x is the test data instance. A detailed derivation of linear score function from Bayes theorem can be found elsewhere (Konishi, 2014; Naik and Kiran, 2019). The linear discriminant score function for the class i (L_i) is given as:

$$L_i = -\frac{1}{2}\vec{\mu}_i'\vec{\Sigma}^{-1}\vec{\mu}_i + \vec{\mu}_i'\vec{\Sigma}^{-1}\vec{x}_i + \log p_i \tag{2.15}$$

where $\vec{\mu}_i$ is the mean vector of each class in the training dataset, $\vec{\Sigma}$ is the covariance matrix of each class in the training dataset, $\vec{x}_i$ is the test data instance, and p_i is the prior probability of class i. Some of the major assumptions of performing LDA are: (i) the training and test data are normally distributed, and (ii) the covariance matrix of each class are equal. However, in reality, the covariance matrix of the classes is not equal, therefore to account for the variances in the dataset, a pooled covariance matrix ($\vec{S}_p$) is considered instead.

$$\vec{S}_p = \frac{\sum_{i=1}^{m}\left(n_i - 1\right)\vec{\Sigma}_i}{\sum_{i=1}^{m}\left(n_i - 1\right)} \tag{2.16}$$

where $\vec{S}_p$ is the pooled covariance matrix for all the classes, m is the number of classes in the dataset, and n_i is the number of observations in each class. Substituting Equation (2.16) in Equation (2.15), the L_i function equation for a class is:

$$L_i = -\frac{1}{2}\vec{\mu}_i'\vec{S}_p^{-1}\vec{\mu}_i + \vec{\mu}_i'\vec{S}_p^{-1}\vec{x}_i + \log p_i \tag{2.17}$$

The L_i is evaluated for all the classes (m) present in the dataset to predict the class of the incoming test data instance ($\vec{x}_i$). The class that yields the maximum L_i score is the class for the test data instance. The relevant processes involved in the algorithm of LDA are presented as pseudocode (Algorithm 2.2).

Algorithm 2.2 The pseudocode of the LDA algorithm

1: **procedure** LDA(training dataset, test data instance)
2:　　let $\vec{x}_i$ = test data instance　　　　　　　　　　▷ data from the test dataset
3:　　**loop**　　　　　　　　　　　　　　　　　　　　　▷ using training dataset
4:　　　　estimate the mean ($\vec{\mu}_i$) vectors for classes　　　　　▷ for 1 to m
5:　　　　determine the covariance matrix ($\vec{\Sigma}$)　　　　　▷ for all m classes
6:　　**end loop**
7:　　**for** (1 to m) **do**　　　　　　　　　　　　　　▷ m - total number of classes
8:　　　　find linear discriminant scores ($L_1, \ldots, L_m$)
9:　　　　estimate linear discriminant function for $\vec{x}_i$　　　▷ using Eq. (1.17); test data
10:　　　　class of $\vec{x}_i$ = max($L_1, \ldots, L_m$)　　　　　▷ class identified; test data
11:　　**end for**
12: **end procedure**

The developed R program demonstrates the working of LDA in classifying the iris dataset (Figure 2.5). The features, such as petal length and width, and sepal length, were selected based on the result of employing the ReleifF algorithm on the iris dataset. The selected influential features were all normally distributed and were used to perform the LDA.

Advantages of LDA:

- Simple, easy to implement, and provides fast classification.
- Provides a linear decision boundary.
- Feature scaling is not required.

Disadvantages of LDA:

- The data should be normally distributed.
- The LDA assumes equal covariance for the classes.

2.3.7.2 k-Nearest Neighbor

The kNN is one of the simplest methods employed for classification applications. This method is also referred to as a lazy or instance-based learner, since like most ML methods, this method does not undergo a training phase before classification.

```
1  # Finding mean of each class ----------------------------------------------
2  cls_1 = subset(trainDat, trainDat$Flower == 1)
3  cls_2 = subset(trainDat, trainDat$Flower == 2)
4  cls_3 = subset(trainDat, trainDat$Flower == 3)
5
6  # Size of each class
7  n1 <- nrow(cls_1); n2 <- nrow(cls_2); n3 <- nrow(cls_3)
8  N = n1 + n2 + n3
9  > print(N)
10 [1] 105
11 n = c(n1, n2, n3)
12 > print(n)
13 [1] 30 40 35
14
15 # Mean of vectors (Sepal length, petal length, and petal width) in each class
16 mu_1 <- rbind(mean(cls_1$SepL), mean(cls_1$PetL), mean(cls_1$PetW))
17 mu_2 <- rbind(mean(cls_2$SepL), mean(cls_2$PetL), mean(cls_2$PetW))
18 mu_3 <- rbind(mean(cls_3$SepL), mean(cls_3$PetL), mean(cls_3$PetW))
19
20 # Finding covariance --------------------------------------------------------
21
22 # Initializing the matrix to store the covariances of each class
23 cov_mat <- array(0, c(3,3)); no_class = 3
24
25 cov1 <- as.data.frame(cov(cls_1)); cov2 <- as.data.frame(cov(cls_2))
26 cov3 <- as.data.frame(cov(cls_3))
27
28 cov_mat[1,1] <- cov1$SepL[1]; cov_mat[1,2] <- cov1$PetL[2]; cov_mat[1,3]<- cov1$PetW[3]
29 cov_mat[2,1] <- cov2$SepL[1]; cov_mat[2,2] <- cov2$PetL[2]; cov_mat[2,3]<- cov2$PetW[3]
30 cov_mat[3,1] <- cov3$SepL[1]; cov_mat[3,2] <- cov3$PetL[2]; cov_mat[3,3]<- cov3$PetW[3]
31
32 # Finding pooled covariance matrix -------------------------------------------
33 sumN = 0; sumD = 0
34 for (i in 1:no_class){
35   sumN = sumN + (n[i]-1)*cov_mat[i,]; sumD = sumD + (n[i]-1)
36   Sp = sumN/sumD
37 }
38 Sp <- diag(Sp)
39
40 # Training data: LDA predictions  --------------------------------------------
41 require(matlib) # for inverse function
42
43 # Creating an array for storing predicted classes ---------------------------
44 ans_max = array(0, c(N, 1))
45 pred_class = array(0, c(N, 1))
46 for (i in 1:N){
47   L1 = (-1/2 * t(mu_1) %*%\% inv(Sp) %*%\% mu_1) + (t(mu_1) \%*\% inv(Sp))
48     \%*\% rbind(trainDat$SepL[i], trainDat$PetL[i], trainDat$PetW[i]) + log(n1/N)
49   L2 = (-1/2 * t(mu_2) %*%\% inv(Sp) %*%\% mu_2) + (t(mu_2) \%*\% inv(Sp))
50     \%*\% rbind(trainDat$SepL[i], trainDat$PetL[i], trainDat$PetW[i]) + log(n2/N)
51   L3 = (-1/2 * t(mu_3) %*%\% inv(Sp) %*%\% mu_3) + (t(mu_3) \%*\% inv(Sp))
52     \%*\% rbind(trainDat$SepL[i], trainDat$PetL[i], trainDat$PetW[i]) + log(n3/N)
53
54   ans_max[i] = max(L1, L2, L3)
55   if (ans_max[i] == L1){
56     pred_class[i, 1] = 1
57   } else if (ans_max[i] == L2){
58     pred_class[i, 1] = 2
59   } else if (ans_max[i] == L3){
60     pred_class[i, 1] = 3
61   }
62 }
63
64 # Training data: Confusion matrix --------------------------------------------
65 require(caret)  # R package for Classification And REgression Training
66 xtab <- table(trainDat$Flower, pred_class)
67 > confusionMatrix(xtab)
68 Confusion Matrix and Statistics
69    pred_class
70      1  2  3
71   1 30  0  0
72   2  0 39  1
73   3  0  3 32
74 Accuracy : 0.9619
75
76 # Test data: Confusion matrix ------------------------------------------------
77 xtab <- table(testDat$Flower, pred_class)
78 > confusionMatrix(xtab)
79 Confusion Matrix and Statistics
80    pred_class
81      1  2  3
82   1 20  0  0
83   2  0 10  0
84   3  0  1 14
85   Accuracy : 0.9778
```

FIGURE 2.5 Manually reconstructed linear discriminant analysis (LDA decoded) for the iris dataset using the linear discriminant score function with evaluation using the confusion matrix.

It delays the modeling of the training dataset unless it is needed to classify the incoming test data instance. The method estimates the similarity between the features in the test and training dataset. Lesser the similarity difference greater chances the attributes belong to the same class. The similarity between the test data instance and every instance in the training dataset is calculated using Euclidean distance, which is the distance between the two points in a plane and which can be calculated using the following equation:

$$D = \sqrt{\sum_{i=1}^{n}(x_i - y_i)^2}$$ (2.18)

where D is the Euclidean distance value between the test and training data instances, n denotes the number of feature attributes in the training dataset, x_i is the training data instance, and y_i is the testing data instance.

The calculated Euclidean distance value is sorted in ascending order, and "k" closest data points are selected from the sorted results. The class of the "k" sorted points that hold the maximum frequency is assigned as the class of the test data instance. The "k" value plays a crucial role in determining the predictive capability of the kNN algorithm. A rule-of-thumb approach for fixing the "k" value is by estimating the square root value of the total training data observations. It is important to note that the predictive performance of the kNN model also depends on the scale of the features. Feature scaling procedures such as normalization or standardization should be performed before employing the kNN algorithm for features existing at different scales (units); this step, however, can be disregarded if the features already have the same units. The pseudocode decoding the kNN algorithm is presented (Algorithm 2.3).

Algorithm 2.3 The pseudocode of the kNN algorithm

1: **procedure** KNN(training dataset, test data instance)
2:     let $\vec{x}_i$ = test data instance   ▷ from the test dataset
3:     **for** (1 to i) **do**   ▷ i - number of observations in testing dataset
4:         **for** (1 to n) **do**   ▷ n - number of observations in training dataset
5:             find Euclidean dist. between test & train instances   ▷ using Eq. (1.18)
6:             for every test data against 'n' in training dataset
7:         **end for**
8:         sort the Euclidean distances in ascending order
9:         select k value   ▷ thumb rule: $k = \sqrt{n}$
10:         pick the k lowest distance (nearest neighbor)   ▷ from data points
11:         find respective classes of the k closest data points
12:         class of $\vec{x}_i$ = maximum occurring class for k closest data points
13:     **end for**
14: **end procedure**

The user-coded algorithm in R demonstrates the working of the kNN method for classification using the selected features with the ReliefF algorithm employing the iris dataset (Figure 2.6). The selected prominent features, petal width and length, and sepal length, had the same units (cm), therefore feature scaling was not performed for this demonstration.

```
 1  # Extracting only the features - iris dataset used  -------------------
 2  trainDat1 <- cbind(trainDat$SepL, trainDat$PetL, trainDat$PetW)
 3  testDat1 <- cbind(testDat$SepL, testDat$PetL, testDat$PetW)
 4
 5  Y <- trainDat$Flower
 6
 7  # Calculating Euclidean distance in 3D space and sorting --------------
 8  Eucld <- array(0, c(nrow(trainDat),1))
 9  cls <- array(0, c(nrow(testDat1),1))
10  for(j in 1:nrow(testDat1)){    # test data
11    for(i in 1: nrow(trainDat1)){    # training data
12      Eucld[i] <- sqrt(((trainDat1[i,1] - testDat1[j,1])^2 +
13                (trainDat1[i,2] - testDat1[j,2])^2 +
14                (trainDat1[i,3] - testDat1[j,3])^2))
15    }
16    srt_dat <- sort(Eucld, index.return = T) # sorting in ascending order
17    k <- round(sqrt(nrow(dat))) # thumb rule
18    srt_dat_ind <- srt_dat$ix[1:k]
19    ypred <- array(0, c(k,1))
20      for(m in 1: k){ # choosing k = 12
21        ypred[m] <- Y[srt_dat_ind[m]]
22      }
23    tab_ypred <- as.data.frame(table(ypred))
24    cls[j] <- ypred[max(tab_ypred$Freq)] # max class frequency in kNN
25  }
26
27  # Confusion matrix ------------------------------------------------------
28  require(caret)
29  xtab <- table(cls, testDat$Flower)
30  > confusionMatrix(xtab)
31
32  Confusion Matrix and Statistics
33  cls  1  2  3
34   1 20  0  0
35   2  0 10  2
36   3  0  0 13
37  Accuracy : 0.9556
```

FIGURE 2.6 Manually reconstructed linear k-nearest neighbor (kNN decoded) algorithm for the iris dataset with evaluation using a confusion matrix.

Advantages of kNN:

- Simple and easy to employ for classification applications since it works only when based on the *k* value and Euclidean distance function.
- The kNN algorithm does not require any training phase because it is an instance-based learner and therefore is faster than the ones that require training such as LDA and naive Bayes.
- The accuracy of the algorithm is not impacted by adding new data to the existing dataset, therefore data can be seamlessly added.

Disadvantages of kNN:

- The kNN does not perform well with large datasets, since it is computationally expensive to calculate the distance between the new test data instance with every other instance in the training dataset.
- Similarly for datasets with high dimensions (many feature attributes) the algorithm becomes more complicated when calculating the distance in each dimension.
- It is sensitive to noisy data, therefore data preprocessing measures should be taken to eliminate any missing values or outliers before employing kNN.

2.3.8 Evaluation of ML Methods

Evaluating the efficiency of the developed model is an essential part of ML. Different metrics are available to evaluate the efficiency of the model, among which confusion matrix, accuracy, precision, recall, and F-score are often used for classification applications.

2.3.8.1 Confusion Matrix

The confusion matrix conveys the correctness of the model in the form of a matrix. It is mostly used for determining the effectiveness of the classification of ML models with two or more classes. The matrix contains the actual values in rows and predicted values in columns (Figure 2.7).

The following are the four essential terms associated with the confusion matrix.

- True positive (TP)—Cases where the actual data belonged to class 1 and the model correctly classified as class 1.
- False negative (FN)—Cases where the actual data belonged to class 1 but the model wrongly classified as class 2.
- False positive (FP)—Cases where the actual data belonged to class 2 but the model wrongly classified as class 1.
- True negative (TN)—Cases where the actual data belonged to class 2 and the model correctly classified as class 2.

The following performance parameters, namely accuracy, precision, recall, F-score, macro-average (based on number of classes), and weighted average (weighting each class by number of samples in each class) are calculated based on the TP, FN, FP, and TN values present in the confusion matrix (Figure 2.7).

2.3.8.2 Accuracy

Accuracy in classification problems is the ratio of the total number of correct predictions by the model (TP and TN) to the total number of input data. It is a good measure for evaluating the model if the classes in the data are nearly balanced.

$$\text{Accuracy} = \left(\frac{TP + TN}{TP + FP + FN + TN}\right) \times 100\% \tag{2.19}$$

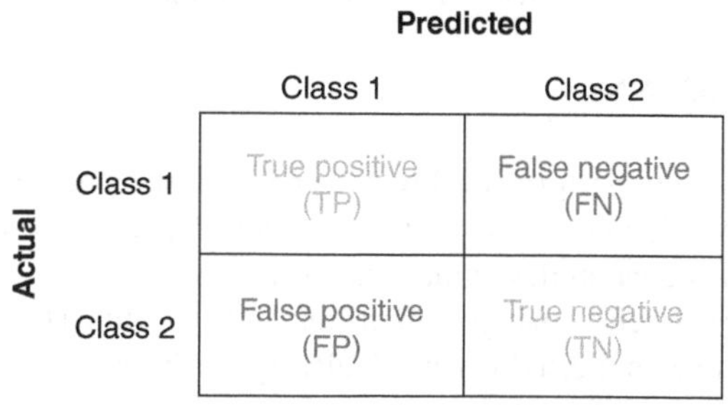

Predicted

		Class 1	Class 2
Actual	Class 1	True positive (TP)	False negative (FN)
	Class 2	False positive (FP)	True negative (TN)

FIGURE 2.7 Confusion matrix for a binary class of data.

2.3.8.3 Precision

Precision is a metric that quantifies the number of cases correctly classified as positive out of the total cases classified as positive. It is the ratio of TP to the sum of TP and FP. The precision result is a value between the range 0.0 and 1.0, where 0.0 represents no precision and 1.0 represents full or perfect precision.

$$\text{Precision} = \left(\frac{TP}{TP + FP} \right) \quad (2.20)$$

2.3.8.4 Recall

Recall is the proportion of cases correctly classified as positive in the class of interest. It is the ratio of TP to the total instances (TP and FN) of a class. It is a value between 0.0 and 1.0.

$$\text{Recall} = \left(\frac{TP}{TP + FN} \right) \quad (2.21)$$

2.3.8.5 F-score

F-score is the balance between the precision and recall and is determined using their harmonic mean. This measure is more reliable than accuracy to validate a model if the data has uneven class distribution. However, with even class distribution (balanced dataset), F-score is the same as the accuracy value.

$$\text{F-score} = 2 \times \left(\frac{\text{Precision} \times \text{Recall}}{\text{Precision} + \text{Recall}} \right) \quad (2.22)$$

2.4 RESULTS AND DISCUSSION

2.4.1 RESULTS OF EVALUATED FEATURES FROM THE DATASET

The evaluated values of the shape features (Equations (2.1)–(2.13)) extracted from digital images of soybean aphid and weed species are presented in Table 2.1. The idea of extracting these shape features is to select a single or combination of shape features that will help in addressing the identification or classification problem among the output classes.

With the soybean aphid dataset having three output classes, some of the features showed good separation between two classes, but not for the third class. Even if a feature showed good separation on the mean value, due to the range of variation (high standard deviations), selecting a single shape feature was not desirable. For example, the mean value of the aspect ratio was distinct for all three classes in the soybean aphid dataset, but the high standard deviation indicates a considerable overlap with other classes. Selecting such a feature will affect the accuracy of the classification algorithm. A similar effect of the aspect ratio can be observed on the weed species dataset as well. Therefore, the selection of suitable features is a

TABLE 2.1

Shape Features Extracted from the Soybean, Aphid, and Weed Species Images Dataset

Shape Feature	Soybean Aphid Dataset (Mean ± SD)			Weed Species Dataset (Mean ± SD)	
	Aphids	Exoskeletons	Leaf Spots	Clover	Grassy
Samples count	109	100	100	120	120
Circularity	0.90 ± 0.06	0.91 ± 0.07	0.65 ± 0.16	0.47 ± 0.15	0.35 ± 0.09
Aspect ratio	1.64 ± 0.21	1.21 ± 0.37	2.01 ± 0.92	3.10 ± 1.35	4.38 ± 1.34
Roundness	0.62 ± 0.08	0.88 ± 0.17	0.56 ± 0.15	0.40 ± 0.20	0.26 ± 0.12
Solidity	0.91 ± 0.02	0.78 ± 0.07	0.83 ± 0.08	0.78 ± 0.08	0.77 ± 0.13
Hollowness	0.09 ± 0.02	0.22 ± 0.07	0.17 ± 0.08	0.22 ± 0.08	0.23 ± 0.13
Reciprocal aspect ratio	0.62 ± 0.08	0.88 ± 0.17	0.56 ± 0.16	0.40 ± 0.20	0.26 ± 0.12
Rectangularity	0.75 ± 0.06	0.63 ± 0.09	0.61 ± 0.12	0.52 ± 0.13	0.40 ± 0.16
Feret major axis ratio	1.05 ± 0.02	1.19 ± 0.07	1.10 ± 0.08	1.04 ± 0.12	1.12 ± 0.17
Area ratio	0.71 ± 0.05	0.62 ± 0.06	0.62 ± 0.08	0.63 ± 0.09	0.57 ± 0.14
Diameter range	4.95 ± 1.86	1.19 ± 1.49	9.51 ± 5.51	26.01 ± 12.94	63.80 + 22.98
Eccentricity	0.78 ± 0.07	0.29 ± 0.34	0.81 ± 0.13	0.88 ± 0.14	0.96 ± 0.06
Area–perimeter ratio	0.07 ± 0.01	0.07 ± 0.01	0.05 ± 0.01	0.04 ± 0.01	0.03 ± 0.01
Feret aspect ratio	1.62 ± 0.19	1.31 ± 0.31	1.85 ± 0.68	2.56 ± 0.98	3.79 ± 1.13

Note: Refer to Figures 2.2 and 2.3 for details.

trial-and-error process, or one can use feature selection algorithms (e.g., ReliefF) to determine the most influential features from the dataset.

2.4.2 SELECTED FEATURES FROM THE DATASET

The datasets with relevant shape features extracted using ImageJ (13 features; Table 2.1) were subjected to the ReliefF algorithm to determine the efficient features influencing the outcome variable. The results showed that shape features, such as (i) roundness, (ii) solidity, and (iii) circularity for soybean aphids; and (i) diameter range, (ii) Feret aspect ratio, and (iii) solidity for the weed species data, emerged as the top three features influencing the outcome; the other two influencing variables with their weights are presented in Table 2.2. The three most influencing variables were used in the original datasets, which were suitably partitioned, and respective models (LDA and kNN) were developed and evaluated using R (Figure 2.1).

2.4.3 DATASET TEST OF NORMALITY FOR MODEL SELECTION

Among the selected ML methods, LDA guarantees minimum classification error when each class in the dataset is normally distributed. Therefore, the accuracy achieved by the LDA model was based on the normality spread of data. The normality test conducted with the Shapiro–Wilk normality test using the R function shapiro.

TABLE 2.2

Feature Ranking Using ReliefF Algorithm Feature Weights for the Two Datasets

Rank	Soybean Aphids (Weights)	Weed Species (Weights)
1	Roundness (0.35)	Diameter range (0.38)
2	Solidity (0.34)	Feret aspect ratio (0.26)
3	Circularity (0.32)	Solidity (0.18)
4	Feret aspect ratio (0.31)	Hollowness (0.17)
5	Diameter range (0.23)	Aspect ratio (0.15)

Note: Refer to Figures 2.2 and 2.3 for details.

test() revealed that the selected features in soybean aphid datasets were normally distributed (H_0: normal; $p > 0.23$, not significant). However, the weed species dataset was not normally distributed (H_0: normal; $p < 3.45 \times 10^{-5}$, significant), therefore LDA is not suitable, while the kNN handles such datasets. Therefore, to yield good model performance, LDA for the soybean aphids and kNN for the weed species dataset were employed for demonstration.

2.4.4 SOYBEAN APHID IDENTIFICATION

2.4.4.1 Features Ranking

The soybean aphid data contained 309 data points (from eight images) of 13 dimensionless shape features extracted and calculated using ImageJ, presented earlier (Equations (2.1)–(2.13)), and three classes in the outcome variable. The dataset was almost balanced with 109, 100, and 100 data points belonging to the classes of aphids, exoskeletons, and leaf spots, respectively. These data points with the three most influential features, namely roundness (0.35), solidity (0.34), and circularity (0.32), which are non-dimensional, formed the original dataset (Table 2.1) and were used in training, developing, and evaluating the LDA ML model.

2.4.4.2 The LDA Model and Evaluation

This dataset was randomly split using the partition ratio 80% and 20% into training (248) and testing (61) data, respectively. The soybean aphid data was observed to be normally distributed and the covariances of the classes were assumed to be equal. The user-coded LDA algorithm (Algorithm 2.2) was successfully trained using the training dataset, and the model was developed. The computation CPU time involved in developing, training, and testing the user-coded LDA algorithm was less than 0.6 s. The confusion matrix (Table 2.3) shows that the trained LDA model was capable of classifying the soybean aphids from the other exoskeletons and leaf spots classes with an overall accuracy (Equation (2.19)) of 92% with the unseen test dataset (Figure 2.8).

In a previous study, using the same soybean aphid images, an identification accuracy of 81% was achieved based on a developed "hollowness"-shape feature to

TABLE 2.3
The LDA Model Evaluation Confusion Matrix and Performance Parameters for Identification of Soybean Aphids, Exoskeletons, and Leaf Spots Using Test Data

Actual/Predicted	Aphids	Exoskeletons	Leaf Spots
Aphids	25	0	1
Exoskeletons	0	14	1
Leaf spots	3	0	17

Performance	Aphids	Exoskeletons	Leaf Spots	Macro-Average	Weighted Average
Support*	28	14	19	61	61
Precision	0.89	1.0	0.90	0.93	0.92
Recall	0.96	0.93	0.85	0.91	0.92
F-score	0.92	0.96	0.87	0.92	0.91
Accuracy	92%			—	—

Note: Refer to Figure 2.8 for details.
* Support is the number of samples in each class.

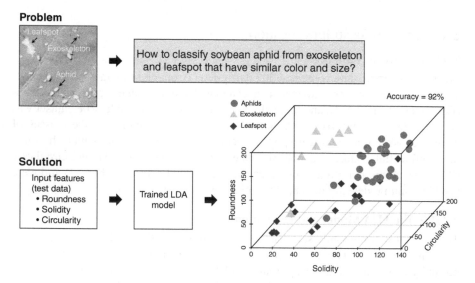

FIGURE 2.8   The LDA classification of soybean aphids, exoskeletons, and leaf spots using the dimensionless shape features, such as solidity, roundness, and circularity.

differentiate aphids from the exoskeletons and leaf spots (Sunoj et al., 2017). Another study developed and based on color and size features to classify soybean aphids achieved an aphid count misclassification ranging from −57% (underestimation) to 142% (overestimation) among different cameras and two lighting conditions (Maharlooei et al., 2017).

The LDA model performance was further evaluated through precision, recall, and F-score, which were determined for each class from the confusion matrix (Table 2.3). The precision (Equation (2.21)) was the highest for exoskeletons at 1.0 followed by leaf spots and aphids at 0.90 and 0.89, respectively. The macro- and weighted averages for the precision were 0.93 and 0.92, respectively. Aphids scored the highest (Equation (2.20)) recall value at 0.96, followed by exoskeletons and leaf spots at 0.93 and 0.85, respectively, while the macro- and weighted average value for recall was 0.91 and 0.92, respectively. The F-score (Equation (2.22)) was the highest for the exoskeletons class (0.96), followed by aphids (0.92) and leaf spots (0.87). The macro-average F-score for the model is 0.92, which is the same as the accuracy achieved (since the classes in the dataset employed are balanced). However, the weighted average for the F-score was slightly higher (0.93) than the macro-average. There was no misclassification between aphids and exoskeletons; however, it was observed between aphids and leaf spots. About 4% of aphids were misclassified as leaf spots, and 15% of leaf spots were misclassified as aphids.

2.4.5 Weed Species Classification

2.4.5.1 Features Ranking

The weed species dataset was well balanced with 13 dimensionless features (Equations (2.1)–(2.13)) and one output variable consisting of two weed species (grassy and clover) classes. The number of data points in the dataset was 240 (from 14 images), with equal data points (120) for the two classes present. The feature selection algorithm ReliefF assigned the highest weights to the feature's diameter range (0.38), Feret aspect ratio (0.26), and solidity (0.18) were considered the most critical features in predicting the class (Table 2.1). Thus, the original dataset consisted of only 240 data points out of these three selected important features; output variables were used in the kNN user-coded model development in R (Algorithm 2.3). The kNN algorithm was selected for the weed species classification application since the dataset was comparatively smaller. The original dataset was split into 70% for training (168 data points) and 30% for testing datasets (72 data points). The split ratio was considered and based on the highest accuracy yielded by the kNN model.

2.4.5.2 The kNN Model and Evaluation

Feature scaling was performed for the kNN model development as the selected critical features scales were in different ranges. Therefore, a scaled testing and training dataset was used in the kNN algorithm. Since there is no training phase, the classification of the weed species through the user-coded kNN algorithm was fast and performed in less than 0.3 s of CPU time. The confusion matrix (Table 2.4) shows that the kNN algorithm successfully achieved the classification of two weed species belonging to the test data with an overall accuracy of 90% (Figure 2.9).

Earlier studies on weed detection in pastureland using a quadratic support vector machine acquired an accuracy of 89% (Zhang et al., 2018). In another study, LDA was used to classify six different species of weeds and soybean and achieved an accuracy of 54% (Gray et al., 2009).

TABLE 2.4
The kNN Model Evaluation Confusion Matrix and Performance Parameters for Identification of Grassy and Clover Weed Species Dataset

Confusion Matrix

Actual/Predicted	Grassy	Clover
Grassy	33	4
Clover	3	32

Performance	Grassy	Clover	Macro-Average	Weighted Average
Support*	36	36	72	72
Precision	0.92	0.88	0.90	0.90
Recall	0.89	0.91	0.90	0.90
F-score	0.90	0.89	0.90	0.90
Accuracy	90%		—	—

Note: Refer to Figure 2.9 for details.
* Support is the number of samples in each class.

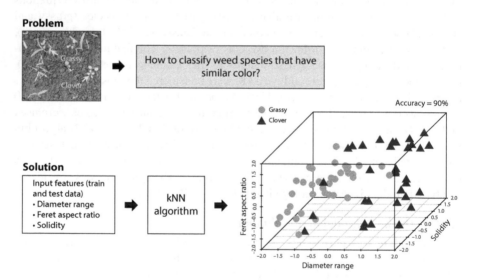

FIGURE 2.9 The kNN classification of grassy and clover weed species using the shape features, such as diameter range, Feret aspect ratio, and solidity.

The precision and F-score (Equations (2.20) and (2.22)) were the highest for the grassy class with values at 0.92 and 0.90, respectively, while the recall (Equation (2.21)) value was the highest for the clover class at 0.91. The macro- and weighted average for precision, recall, and F-score were the same at 0.90. Misclassification was observed in both the weed species; however, it was less than 10%. As the model can be readily applied to other and similar images, analysis with more sample images will provide a better estimate of the accuracy and misclassification values.

2.4.6 COMPARISON OF RESULTS WITH THE STANDARD IRIS DATA

The performance of ML models (LDA and kNN) with user-coded algorithms (Algorithms 2.2 and 2.3) and the corresponding "black-box" methods (readily available R packages) estimated for the iris, soybean aphids, and weed species datasets produced comparable results (Table 2.5). Between the accuracy and F-score (calculated using recall and precision) performance parameters, accuracy was selected for comparison since all the three datasets considered were balanced with even class distribution. In addition to accuracy, the computation process time was also considered for comparing the performance of the models and methods. The R function train() from the caret package was used to conduct the LDA and kNN model development. The LDA and kNN methods employed for the standard iris dataset produced a good performance with 98% and 96% accuracies with the user-coded algorithm, and 100% and 93% with the black-box method, respectively. The LDA performed slightly better than kNN for the iris dataset since the data points were normally distributed ($p > 0.08$).

Although LDA was considered the best ML method for the soybean aphid dataset owing to its normality spread, kNN was also employed for comparison, and, as expected, LDA performed better (92%) than kNN (84%) with both the user-coded algorithm and black-box method. The accuracy of both methods coincided with the soybean aphid dataset. Similarly, in addition to the kNN method employed for the weed species dataset with no normality, the LDA method was also tested for performance. As expected, the kNN performed better (90% user-coded, 92% black-box) than LDA (81% user-coded, 79% black-box). The user-coded algorithm for LDA produced better or comparable accuracy for the soybean aphid and weed species datasets, while the user-coded kNN produced better or comparable accuracy for iris and soybean aphid datasets.

TABLE 2.5

Comparison of ML User-Coded and Blackbox Methods' Performance with the Standard Iris and Soybean Aphid and Weed Species Datasets Based on Accuracy and CPU Time

	User-Coded Algorithm in R:		Black-Box Code in R:	
	Accuracy (CPU Time, s)*		Accuracy (CPU Time, s)*	
Dataset	LDA	kNN	LDA	kNN
Iris	98% (0.46 s)	96% (0.10 s)	100% (0.72 s)	93% (2.77 s)
Soybean aphids	92% (0.60 s)	84% (0.23 s)	92% (0.70 s)	84% (3.05 s)
Weed species	81% (0.39 s)	90% (0.22 s)	79% (0.81 s)	92% (2.94 s)

Note: Other performance parameters were not considered for comparison. CPU—central processing unit.
* Mean of ten runs of the models, not including data visualization.

Another significant performance parameter of the user-coded algorithms and black-box methods is the computation process time measured in terms of the central processing unit (CPU) time taken in seconds, which was recorded using the R Sys. time() function. By appropriate placement of the system's time function before and after the code blocks to capture the start and end times, the CPU times of specific code blocks, such as developing, training, or evaluating, or whole method processing can be evaluated. The computation time for kNN, an instance-based learner, was always less by an average of 42% than LDA for the user-coded algorithm since kNN has no training phase. However, this trend of reduced CPU time was observed only for user-coded algorithms, while for the black-box method the CPU time for kNN was consistently higher than LDA with an average of a 292% increase. Also, it can be observed that the CPU times of the user-coded algorithms were always lower by an average of 35% for LDA and 94% for kNN than the black-box method. This interesting result may be due to the direct approach by the user-coded methods compared to the overhead associated with the black-box methods. This improved performance of the user-coded methods will have an even greater impact with big datasets and supports the case for developing customized user-coded algorithms over the black-box approaches.

2.5 CONCLUSIONS

This chapter has successfully demonstrated the decoding and reconstructing of the common ML classification methods, such as LDA and kNN, from the underlying mathematics behind these algorithms, applied to two agricultural applications (soybean aphids and weed species) using open source platforms like ImageJ and R through developed user-coded programs. The decoded ReliefF algorithm in each dataset (soybean aphids, weed species, and iris—for comparison) selected the most influential features, among 13 shape-based features, for developing the reconstructed classification algorithms. Features such as roundness, solidity, and circularity for soybean aphids; diameter range, Feret aspect ratio, and solidity for weed species; and sepal length, petal length, and petal width for iris; emerged as the best based on weights.

The prediction accuracy in classification for reconstructed user-coded and direct black-box approaches were comparable for both LDA and kNN models. The LDA was more accurate in classifying the soybean aphids (92%) and iris (≥98%), while kNN performed better for the weed species (≥90%) dataset classes based on the normality distribution of the data. The averages (macro and weighted) of performance parameters, such as precision, recall, and F-scores, were ≥0.90, indicating the good performance of the user-coded models.

The computation time for the kNN was less than the LDA model (by 42%) for the user-coded approach, but the LDA was less than the kNN model (by 75%) while using the black-box approach. Overall, the computational time for user-coded algorithms was decreased by 35% and 94% for LDA and kNN, respectively, compared to the black-box, and this represents a significant impact on ML applications development, especially while handling large datasets.

In addition to the advantage of cutting down the computational time and expense in turn, understanding the logic behind the ML algorithm is central as it allows the user to better understand the methods, aids in taking better post-hoc decisions in the case of prediction failure of a model, and develops customized application-specific methodologies. However, some of the limitations of decoding the ML algorithm includes (i) knowledge of subjects such as statistics and mathematics to comprehend the theory behind the ML methods; (ii) coding skills to construct an ML algorithm; and (iii) time and effort required to understand and build models.

With a better knowledge of the theories and fundamental working principles, in the future novel ML algorithms can be developed; also any existing machine learning algorithm can be tweaked to meet the specific needs of the data in hand. Therefore, similar to LDA and kNN dealt with in this chapter, the mathematics behind all the other available ML methods can be decoded and reconstructed using free and rich open source resources, thereby breaking the "black-boxes" and making user-coded "efficient tools" for ML applications.

ACKNOWLEDGMENTS

The authors are grateful to the collaboration extended by Drs. Jose G. Franco, John Hendrickson, and David Archer, Northern Great Plains Research Laboratory (NGPRL), USDA-ARS, Mandan, ND and for the weed species images. This work was supported in part by the NGPRL, Grant number: FAR0028541, and the USDA National Institute of Food and Agriculture, Hatch Project: ND01481, Accession number: 1014700. This support is greatly appreciated.

REFERENCES

Abadi, M., A. Agarwal, P. Barham, E. Brevdo, Z. Chen, C. Citro, G. S. Corrado, A. Davis, J. Dean, M. Devin, S. Ghemawat, I. Goodfellow, A. Harp, G. Irving, M. Isard, Y. Jia, R. Jozefowicz, L. Kaiser, M. Kudlur, J. Levenberg, D. Mané, R. Monga, S. Moore, D. Murray, C. Olah, M. Schuster, J. Shlens, B. Steiner, I. Sutskever, K. Talwar, P. Tucker, V. Vanhoucke, V. Vasudevan, F. Viégas, O. Vinyals, P. Warden, M. Wattenberg, M. Wicke, Y. Yu, and X. Zheng. 2015. TensorFlow: Large-scale machine learning on heterogeneous systems. Available at: https://www.tensorflow.org.

Ali Alheeti, K. M. and K. McDonald-Maier. 2018. Intelligent intrusion detection in external communication systems for autonomous vehicles. *Systems Science & Control Engineering* 6(1):48–56.

Bai, S., X. Zhou, Y. Lyu, J. Wang, and C. Pan. 2019. Data visualization model methods and techniques. In *IOP Conference Series: Earth and Environmental Science*, Volume 252, pp. 052063-1–052063-10. London, UK: IOP Publishing.

Behmann, J., A.-K. Mahlein, T. Rumpf, C. Römer, and L. Plümer. 2015. A review of advanced machine learning methods for the detection of biotic stress in precision crop protection. *Precision Agriculture* 16(3):239–260.

Borchers, M. R., Y. M. Chang, K. L. Proudfoot, B. A. Wadsworth, A. E. Stone, and J. M. Bewley. 2017. Machine-learning-based calving prediction from activity, lying, and ruminating behaviors in dairy cattle. *Journal of Dairy Science* 100(7):5664–5674.

Choi, S. Y. and D. Cha. 2019. Unmanned aerial vehicles using machine learning for autonomous flight; state-of-the-art. *Advanced Robotics* 33(6):265–277.

Chu, X., I. F. Ilyas, S. Krishnan, and J. Wang. 2016. Data cleaning: Overview and emerging challenges. In *Proceedings of the 2016 International Conference on Management of Data*, pp. 2201–2206. New York: ACM.

Das, H., A. K. Jena, J. Nayak, B. Naik, and H. Behera. 2015. A novel PSO based back propagation learning-MLP (PSO-BP-MLP) for classification. In *Computational Intelligence in Data Mining*, Volume 2, pp. 461–471. Vienna: Springer.

Das, H., B. Naik, and H. Behera. 2018. Classification of diabetes mellitus disease (DMD): A data mining (DM) approach. In *Progress in Computing, Analytics and Networking*, pp. 539–549. Singapore: Springer.

Das, H., B. Naik, and H. Behera. 2020a. A hybrid neuro-fuzzy and feature reduction model for classification. *Advances in Fuzzy Systems* 2020:1–15.

Das, H., B. Naik, and H. Behera. 2020b. Medical disease analysis using neuro-fuzzy with feature extraction model for classification. *Informatics in Medicine Unlocked* 18:100288.

Das, H., B. Naik, H. Behera, S. Jaiswal, P. Mahato, and M. Rout. 2020. Biomedical data analysis using neuro-fuzzy model with post-feature reduction. *Journal of King Saud University-Computer and Information Sciences*. (DOI: https://doi.org/10.1016/j.jksuci.2020.01.007. Article in press).

Das, H., B. Naik, and H. S. Behera. 2020c. An experimental analysis of machine learning classification algorithms on biomedical data. In S. Kundu, U. S. Acharya, C. K. De, and S. Mukherjee (Eds.), *Proceedings of the 2nd International Conference on Communication, Devices and Computing*, pp. 525–539. Singapore: Springer.

Dey, N., H. Das, B. Naik, and H. S. Behera. 2019. *Big Data Analytics for Intelligent Healthcare Management*. Boston, MA: Academic Press.

dos Santos Ferreira, A., D. M. Freitas, G. G. da Silva, H. Pistori, and M. T. Folhes. 2017. Weed detection in soybean crops using convnets. *Computers and Electronics in Agriculture* 143:314–324.

Du, C.-J. and D.-W. Sun. 2004. Recent developments in the applications of image processing techniques for food quality evaluation. *Trends in Food Science & Technology* 15(5):230–249.

Endelman, J. B. 2011. Ridge regression and other kernels for genomic selection with R package rrBLUP. *The Plant Genome* 4(3):250–255.

Ferentinos, K. P. 2018. Deep learning models for plant disease detection and diagnosis. *Computers and Electronics in Agriculture* 145:311–318.

Fisher, R. A. 1936. The use of multiple measurements in taxonomic problems. *Annals of Eugenics* 7(2):179–188.

Gomez-Uribe, C. A. and N. Hunt. 2016. The netflix recommender system: Algorithms, business value, and innovation. *ACM Transactions on Management Information Systems (TMIS)* 6(4):13.

Gray, C. J., D. R. Shaw, and L. M. Bruce. 2009. Utility of hyperspectral reflectance for differentiating soybean (*Glycine max*) and six weed species. *Weed Technology* 23(1):108–119.

Igathinathane, C., L. O. Pordesimo, E. P. Columbus, W. D. Batchelor, and S. R. Methuku. 2008. Shape identification and particles size distribution from basic shape parameters using ImageJ. *Computers and Electronics in Agriculture* 63(2):168–182.

Igathinathane, C., L. O. Pordesimo, E. P. Columbus, W. D. Batchelor, and S. Sokhansanj. 2009. Sieveless particle size distribution analysis of particulate materials through computer vision. *Computers and Electronics in Agriculture* 66(2):147–158.

Jin, X., A. Xu, R. Bie, and P. Guo. 2006. Machine learning techniques and chi-square feature selection for cancer classification using sage gene expression profiles. In *International Workshop on Data Mining for Biomedical Applications*, pp. 106–115. Berlin: Springer.

Joachims, T. 1998. Text categorization with support vector machines: Learning with many relevant features. In *European Conference on Machine Learning*, pp. 137–142. Springer.

Kamarulzalis, A. H. and M. A. A. Abdullah. 2019. An improvement algoithm for iris classification by using linear support vector machine (LSVM), k-Nearest Neighbours (k-NN) and Random Nearest Neighbous (RNN). *Journal of Mathematics & Computing Science* 5(1):32–38.

Kira, K. and L. A. Rendell. 1992. A practical approach to feature selection. In *Machine Learning Proceedings 1992*, pp. 249–256. Burlington: Elsevier.

Konishi, S. 2014. *Introduction to Multivariate Analysis: Linear and Nonlinear Modeling.* Hoboken, NJ: CRC Press.

Kononenko, I. 1994. Estimating attributes: Analysis and extensions of RELIEF. In *European Conference on Machine Learning*, pp. 171–182. Berlin: Springer.

Kuhn, M. 2019. caret: Classification and Regression Training. R package version 6.0-84.

Kültür, Y. and M. U. Çağlayan. 2017. A novel cardholder behavior model for detecting credit card fraud. *Intelligent Automation & Soft Computing*:1–11.

Li, F., Y. Yang, and E. P. Xing. 2006. From lasso regression to feature vector machine. In *Advances in Neural Information Processing Systems*, pp. 779–786. Cambridge, MA: MIT Press.

Liakos, K. G., P. Busato, D. Moshou, S. Pearson, and D. Bochtis. 2018. Machine learning in agriculture: A review. *Sensors* 18(8):2674.

Löw, F., C. Biradar, O. Dubovyk, E. Fliemann, A. Akramkhanov, A. Narvaez Vallejo, and F. Waldner. 2018. Regional-scale monitoring of cropland intensity and productivity with multi-source satellite image time series. *GIScience & Remote Sensing* 55(4):539–567.

Maharlooei, M., S. Sivarajan, S. G. Bajwa, J. P. Harmon, and J. Nowatzki. 2017. Detection of soybean aphids in a greenhouse using an image processing technique. *Computers and Electronics in Agriculture* 132:63–70.

Mao, K. Z. 2004. Orthogonal forward selection and backward elimination algorithms for feature subset selection. *IEEE Transactions on Systems, Man, and Cybernetics, Part B (Cybernetics)* 34(1):629–634.

Martinez, W. L. and A. R. Martinez. 2015. *Computational Statistics Handbook with MATLAB.* Boca Raton: Chapman and Hall/CRC.

Naik, D. L. and R. Kiran. 2019. Identification and characterization of fracture in metals using machine learning based texture recognition algorithms. *Engineering Fracture Mechanics* 219:106618.

Pedregosa, F., G. Varoquaux, A. Gramfort, V. Michel, B. Thirion, O. Grisel, M. Blondel, P. Prettenhofer, R. Weiss, V. Dubourg, J. Vanderplas, A. Passos, D. Cournapeau, M. Brucher, M. Perrot, and E. Duchesnay. 2011. Scikit-learn: Machine learning in Python. *Journal of Machine Learning Research* 12:2825–2830.

R Core Team. 2019. R: A language and environment for statistical computing. Available at: https://www.R-project.org/.

Ramasubramanian, K. and A. Singh. 2017. *Machine Learning Using R.* Springer.

Rao, C. R. 1948. The utilization of multiple measurements in problems of biological classification. *Journal of the Royal Statistical Society. Series B (Methodological)* 10(2):159–203.

Rasband, W. S. 2019. ImageJ, US National Institutes of Health, Bethesda, MD. Available at: https://imagej.nih.gov/ij/.

Reitermanova, Z. 2010. Data splitting. In *WDS*, Volume 10, pp. 31–36.

Robnik-Šikonja, M. and I. Kononenko. 2003. Theoretical and empirical analysis of ReliefF and RReliefF. *Machine Learning* 53(1-2):23–69.

Sarkar, D., R. Bali, and T. Sharma. 2018. Practical machine learning with Python. In *A Problem-Solvers Guide to Building Real-World Intelligent Systems*, Berkely, CA.

Schindelin, J., I. Arganda-Carreras, E. Frise, V. Kaynig, M. Longair, T. Pietzsch, S. Preibisch, C. Rueden, S. Saalfeld, B. Schmid, et al. 2012. Fiji: an open-source platform for biological-image analysis. *Nature Methods* 9(7):676–682.

Singh, N. and V. Srivastava. 2019. Iris data classification using modified fuzzy C means. In *Computational Intelligence: Theories, Applications and Future Directions*, Volume *I*, pp. 345–357. Singapore: Springer.

Sunoj, S., S. Sivarajan, M. Maharlooei, S. G. Bajwa, J. P. Harmon, J. F. Nowatzki, and C. Igathinathane. 2017. Identification and counting of soybean aphids from digital images using shape classification. *Transactions of the ASABE* 60(5):1467–1477.

Sunoj, S., S. Subhashree, S. Dharani, C. Igathinathane, J. Franco, R. Mallinger, J. Prasifka, and D. Archer. 2018. Sunflower floral dimension measurements using digital image processing. *Computers and Electronics in Agriculture* 151:403–415.

Tantalaki, N., S. Souravlas, and M. Roumeliotis. 2019. Data-driven decision making in precision agriculture: The rise of big data in agricultural systems. *Journal of Agricultural & Food Information* 20(4):344–380.

Urbanowicz, R. J., M. Meeker, W. La Cava, R. S. Olson, and J. H. Moore. 2018. Relief-based feature selection: Introduction and review. *Journal of Biomedical Informatics* 85:189–203.

Wang, A., W. Zhang, and X. Wei. 2019. A review on weed detection using ground-based machine vision and image processing techniques. *Computers and Electronics in Agriculture* 158:226–240.

Yen, J.-C., F.-J. Chang, and S. Chang. 1995. A new criterion for automatic multilevel thresholding. *IEEE Transactions on Image Processing* 4(3):370–378.

Zhang, W., M. F. Hansen, T. N. Volonakis, M. Smith, L. Smith, J. Wilson, G. Ralston, L. Broadbent, and G. Wright. 2018. Broad-leaf weed detection in pasture. In *2018 IEEE 3rd International Conference on Image, Vision and Computing (ICIVC)*, pp. 101–105. Piscataway, NJ: IEEE.

Zhou, C., H. Ye, Z. Xu, J. Hu, X. Shi, S. Hua, J. Yue, and G. Yang. 2019. Estimating maize-leaf coverage in field conditions by applying a machine learning algorithm to UAV remote sensing images. *Applied Sciences* 9(11):2389.

Zou, H. and T. Hastie. 2005. Regularization and variable selection via the elastic net. *Journal of the Royal Statistical Society: Series B (Statistical Methodology)* 67(2):301–320.

3 A Multi-Stage Hybrid Model for Odia Compound Character Recognition

Dibyasundar Das, Deepak Ranjan Nayak, Ratnakar Dash, and Banshidhar Majhi

CONTENTS

3.1 INTRODUCTION

Optical character recognition (OCR) for Indic languages has recently gained increasing popularity due to its high usage in many real-time applications. Most of the earlier works have shown higher recognition performance for languages like English, Japanese, and Chinese. However, the performance for Indic languages is still far away from real-time demands. Odia is one of the constitutional languages of India, specifically used in the states of Odisha, Jharkhand, and West Bengal. There are about 27 million people who use Odia as their basic language of communication. The major challenge in modeling an Odia OCR is the detection of compound characters with higher accuracy. Odia script contains more than 400 character classes that include vowels, consonants, matras (vowel allographs), compound characters (consonant allographs), some special symbols and numerals. The proposed method divides the character recognition task into 211 classes of characters and detects the class label using a three-stage hybrid architecture. It is worth mentioning here that the method is applied only to Odia printed characters.

OCR development has a rich history [8, 30, 34, 37]. Starting from OCR machine to modern digital software-based recognition systems, the main aim is to recognize and classify characters irrespective of position, font size, and their orientation in the visual field. Now with modern devices, making the detection font style independent has also become an interesting field of research. Broadly an OCR system has two steps, namely, character extraction and character detection. The extraction process involves image enhancement [1], skew and error correction, binarization [18, 19, 42–44], line and word separation, and character separation. The image enhancement and error removal techniques aim to make the image ready for the OCR engine. These steps are generally referred to as the pre-processing step as a whole, which can be followed by a font detection process [31]. Font detection helps to improve accuracy by utilizing font-specific detection algorithms. So far little work has been reported in relation to printed Odia character recognition, and in most of these cases only primary characters [7, 29] have been considered. The study of basic and compound characters in many of the Indian languages like Bangla [6, 11] and Devnagari [34] has been carried out by many researchers. Chaudhuri et al. [6] have studied 75 basic and modified characters of the Bangla language and achieved 96% accuracy. Following this, 300 numbers of allographs of the Bangla script were studied by Garain et al. [17] using the run-number-based hybrid method and showed 99.69% of accuracy over a self-collected dataset, while in [34], an accuracy of 96.00% was earned for Devanagari character recognition. However, the performance is yet to be improved significantly for other Indian scripts like Kannada [3], Gurumukhi [21], Telugu [46], Tamil [9], Gujarati [2, 16], and Odia [7, 29].

Many studies have been carried out for handwritten Odia OCR. Padhi et al. [32, 33] designed features based on zone segmentation and its statistical information, and used ANN to classify 49 allographs of Odia script, with an accuracy of 94% achieved on a self-collected handwritten character dataset. Pal et al. [36] studied 52 Odia characters using gradient and curvature features with a quadratic classifier to achieve an accuracy of 91.11%. Kumar et al. [20] proposed a model using an Ant miner algorithm on 50 classes of Odia handwritten characters and achieved 90% accuracy. Basa et al. [4, 25] designed a two-stage classification system. They proposed a tree based method to classify Odia handwritten characters into two groups and further used discriminant features of each group learned by two different ANNs. They formed a handwritten Odia character dataset of 51 classes and showed that their method provided an overall accuracy of 90%. Das et al. [10] proposed a recognition model for Odia characters using an extreme learning machine (ELM). They analyzed various parameters of ELM and their effect on accuracy for various datasets.

Similarly, many works have been published on handwritten Odia numerals which are summarized as follows. One of the earliest works on handwritten Odia numeral recognition is reported by Sarangi et al. [41] where a Hopfield neural network is used to achieve 95.4% accuracy. Following this, Sarangi et al. [39] proposed a recognition model by use of a lower and upper triangular matrix (LU) feature with a multi-layer perceptron network, which gives 85.3% accuracy. Bhowmik et al. [5] reported recognition of Odia numerals with a hidden Markov model and the occurrence of strokes in characters. They obtained an accuracy of 90.5% from the experiment. Roy et al. [38] studied the histogram block-based feature for handwritten numeral recognition and achieved an accuracy of 94.81%. Sarangi et al. [40] obtained 85.3% accuracy using LU factorization as a feature extractor and a Naive Bayes classifier. Dash et al. [12] used a curvature feature and Kirsh edge operator with two well-known classifiers such as a discriminative-learning quadratic-discriminant function and a modified quadratic-discriminant function to achieve an accuracy of 98.4% and 98.5% respectively. Mishra et al. [26] investigated DCT and DWT features independently to achieve 87.5% and 92% accuracy respectively. Dash et al. [13, 14] proposed an Odia numeral recognition system using a zone-based non-redundant Socketwell transform and a k-nearest neighbor (k-NN) classifier to obtain an accuracy of 98.80%. Mahto et al. [24] designed a method where a simple ANN with a Quadrant-mean-based feature could achieve a 93.2% accuracy. Mishra et al. [28] used a contour-based feature with HMM to yield 96.3% accuracy on an Odia numeral dataset. Mishra et al. [27] employed using cord length and an angle feature for a handwritten Odia numeral recognition model. Dash et al. [15] used binary external-symmetry-axis-based features and classified a self-collected handwritten numeral dataset with an accuracy of 95%.

From this literature survey, it is evident that most of the research is limited to the detection of only 51 basic Odia characters and ten numeric characters. However, Odia has more than 400 classes of characters which are generated by combining one or more of them. Some of the sample characters and their category are given in Figure 3.1. Further, Most of the aforementioned works are dedicated in detecting 11 vowels,

Name	Characters
Vowels	ଅ, ଆ, ଇ, ଈ, ଉ, ଊ, ଋ, ଏ, ଐ, ଓ, ଔ
Consonants	କ, ଖ, ଗ, ଘ, ଙ, ଚ, ଛ, ଜ, ଝ, ଞ, ଟ, ଠ, ଡ, ଢ, ଣ, ତ, ଥ, ଦ, ଧ, ନ, ପ, ଫ, ବ, ଭ, ମ, ଯ, ର, ଲ, ଶ, ଷ, ସ, ହ, ୟ, ଳ
Numbers	୦, ୧, ୨, ୩, ୪, ୫, ୬, ୭, ୮, ୯
Special characters	anusar (◌ଂ), chandrabindu (◌ଁ), bhisarga (◌ଃ)
Matras (vowel modifiers)	◌ା, ◌ି, ◌ୀ, ◌ୁ, ◌ୂ, ◌ୃ, ◌େ, ◌ୈ, ◌ୋ, ◌ୌ
Two-character conjucts	ଙ୍କ, ଙ୍ଖ, ଙ୍ଗ, ଙ୍ଘ, ଞ୍ଚ, ଞ୍ଛ, ଞ୍ଜ, ଞ୍ଝ, ଣ୍ଟ, ଣ୍ଠ, ଣ୍ଡ, ଣ୍ଢ, ନ୍ତ, ନ୍ଥ, ନ୍ଦ, ନ୍ଧ, ମ୍ପ, ମ୍ଫ, ମ୍ବ, ମ୍ଭ, କ୍କ, କ୍ଟ, କ୍ତ, କ୍ବ, କ୍ମ, କ୍ର, କ୍ଲ, କ୍ସ, କ୍ଷ, କ୍ୟ, କ୍ଳ, ଖ୍ୟ, ଗ୍ଧ, ଗ୍ନ, ଗ୍ବ, ଗ୍ଲ, ଗ୍ଳ, ଗ୍ମ, ଗ୍ର, ଘ୍ୟ, ଚ୍ଚ, ଚ୍ଛ, ଚ୍ୟ, ଜ୍ଜ, ଜ୍ଞ, ଜ୍ଝ, ଜ୍ୟ, ଜ୍ର, ଜ୍ବ, ଟ୍ଟ, ଟ୍ବ, ଟ୍ର, ଟ୍ୟ, ଠ୍ୟ, ଡ୍ଡ, ଡ୍ଢ, ଡ୍ର, ଡ୍ୟ, ଢ୍ୟ, ଣ୍ଣ, ତ୍ତ, ତ୍ଥ, ତ୍ନ, ତ୍ପ, ତ୍ବ, ତ୍ମ, ତ୍ର, ତ୍ସ, ତ୍ୟ, ଥ୍ୟ, ଦ୍ଗ, ଦ୍ଦ, ଦ୍ଧ, ଦ୍ବ, ଦ୍ଭ, ଦ୍ମ, ଦ୍ର, ଦ୍ୟ, ଧ୍ବ, ଧ୍ର, ଧ୍ମ, ନ୍ନ, ନ୍ମ, ନ୍ୟ, ପ୍ଟ, ପ୍ତ, ପ୍ନ, ପ୍ପ, ପ୍ର, ପ୍ଲ, ପ୍ସ, ପ୍ୟ, ଫ୍ଲ, ଫ୍ର, ଫ୍ଳ, ବ୍ଦ, ବ୍ଜ, ବ୍ଧ, ବ୍ର, ବ୍ୟ, ବ୍ଲ, ଭ୍ର, ଭ୍ୟ, ମ୍ନ, ମ୍ମ, ମ୍ର, ମ୍ଲ, ମ୍ୟ, ଯ୍ୟ, ର୍କ, ର୍ଖ, ର୍ଗ, ର୍ଘ, ର୍ଚ, ର୍ଜ, ର୍ଝ, ର୍ଣ, ର୍ତ, ର୍ଥ, ର୍ଦ, ର୍ଧ, ର୍ପ, ର୍ଫ, ର୍ବ, ର୍ମ, ର୍ଶ, ର୍ଯ, ର୍ଷ, ଲ୍ବ, ଲ୍କ, ଲ୍ମ, ଶ୍ଚ, ଶ୍ଛ, ଶ୍ନ, ଶ୍ବ, ଶ୍ମ, ଶ୍ର, ଶ୍ଲ, ଶ୍ୟ, ଶ୍ଳ, ଷ୍କ, ଷ୍ଟ, ଷ୍ଠ, ଷ୍ଣ, ଷ୍ପ, ଷ୍ଫ, ଷ୍ୟ, ସ୍କ, ସ୍ଖ, ସ୍ତ, ସ୍ଥ, ସ୍ନ, ସ୍ପ, ସ୍ଫ, ସ୍ବ, ସ୍ମ, ସ୍ର, ସ୍ୟ, ସ୍ଲ, ହ୍ନ, ହ୍ବ, ହ୍ମ, ହ୍ର, ହ୍ଲ, ହ୍ୟ, କ୍ଷ୍ଣ, କ୍ଷ୍ମ, କ୍ଷ୍ୟ, ଲ୍କ, ଲ୍ପ, ଲ୍ୟ, ଲ୍ଲ
Three-character conjucts	କ୍ତ୍ର, ଙ୍କ୍ଷ, ଙ୍କ୍ତ, ଙ୍ଗ୍ର, ଣ୍ଡ୍ର, ତ୍ତ୍ବ, ତ୍ତ୍ର, ତ୍ମ୍ୟ, ସ୍ତ୍ର, ସ୍ଥ୍ୟ, ନ୍ତ୍ୟ, ନ୍ତ୍ର, ନ୍ଦ୍ବ, ନ୍ଦ୍ର, ନ୍ଧ୍ବ, ନ୍ଧ୍ର, ନ୍ଧ୍ୟ, ନ୍ତ୍ୟ, ମ୍ପ୍ର, ନ୍ତ୍ର୍, ର୍ଚ୍ଚ, ର୍ଚ୍ଛ, ର୍ଣ୍ଡ, ର୍ତ୍ତ, ର୍ଦ୍ଧ, ର୍ଦ୍ବ, ର୍ଷ୍ଣ, ର୍ଦ୍ର, ର୍ଯ୍ୟ, ର୍ଶ୍ବ, ଷ୍ଟ୍ର, ଷ୍ଣ୍ବ, ସ୍ତ୍ର, ସ୍ତ୍ବ, ସ୍ଥ୍ୟ, ନ୍ତ୍ୟ, ର୍ତ୍ୟ, ର୍ଲ୍ଲ, ର୍ତ୍ତ୍ର, ର୍ଥ୍ୟ, ର୍ଘ୍ୟ, ର୍ତ୍ୟ, ଙ୍କ୍ଷ, ଷ୍ଟ୍ର, ସ୍କ୍ର

FIGURE 3.1 Allograph in Odia script.

35 consonants, and 10 numeric values. However, the synthesis of most Odia sentences is not possible without the use of martas and compound characters. Odia OCR needs to detect basic as well as compound characters to convert Odia scanned documents to their corresponding text. We studied the various literature and found that 211 classes of characters are ample for recognizing most of the documents. Our research focused on these 211 characters. The current models are not efficient enough to recognize all these characters, as shown in Section 3.5. Hence, aiming at improving recognition performance, we proposed a hybrid model which utilizes a structural similarity index (SSIM) in the first stage, a projection profile and Kendall rank correlation coefficient ranking in the second stage, and a local frequency descriptor (LFD) and General regression neural network (GRNN) in the final stage to predict the final class label. Different combinations of these methods were tested over a newly created dataset and a comparative analysis was made with state- of-the-art methods. An overall accuracy of 90.6% was achieved using the proposed three-stage hybrid model.

The chapter is organized in the following way. Section 3.2 gives a description of the methods that are required to build the hybrid model. Section 3.3 details the proposed model and Section 3.4 describes the dataset developed and the experimental details. Following this, in Section 3.5, results of this investigation are presented, and finally Section 3.6 concludes the chapter.

3.2 BACKGROUND

3.2.1 General OCR Stages

Generally, a complete OCR process is divided into multiple stages. Image acquisition is the first and essential step for any computer vision application. For character recognition images are captured by a flatbed scanner. A digital image of the document can also be obtained with powerful mobile cameras, though they need additional processing. After image acquisition, the known errors are to be removed by the pre-processing step. Image enhancement, noise removal, skew detection and correction, page segmentation, binarization, and font detection are examples of a few pre-processing methods that are often used in OCR models. Segmentation is the process of separating lines, words, and characters and has higher significance in determining the recognition accuracy in the case of handwritten and touched character recognition. Each separated character is fed into the recognition model to determine the class label. Models can be derived by various learning and classification methods and can be divided broadly into the two categories of template matching and feature matching hybrid models (feature + template). After recognizing the character class, its corresponding ASCII or UNI-code must be written into a text file. Handling both word and line separation is essential in this stage. The information obtained from the segmentation step is used in this context, though the recognition process may not lead to the desired result. The presence of punctuation and errors may also result in some unnecessary characters, which can be solved with post-dictionary matching and contextual information. An overview of the general model is shown in Figure 3.2.

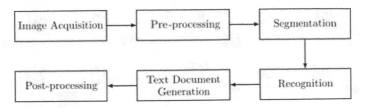

FIGURE 3.2 An overview of the OCR process.

3.2.2 STRUCTURAL SIMILARITY

Human eyes focus on the structure of an image rather than its illumination. The SSIM [47] exploits this principle and uses luminance, contrast, and structure comparison to give ranking value to image quality that is very similar to a subjective human evaluation standard. This is expressed mathematically in Equations 3.1 and 3.2. The overall presentation of the SSIM evaluation is shown in Figure 3.3.

$$\text{SSIM} = \left[l(x, y)\right]^{\alpha} \left[c(x, y)\right]^{\beta} \left[s(x, y)\right]^{\gamma} \tag{3.1}$$

where s denotes structure, c denotes contrast, and l indicates local luminance. The expression correlates to the human visual system and by taking $\alpha = \beta = \gamma = 1$ it can be simplified as

$$\text{SSIM} = \frac{\left(2\mu_{aax}\mu_y + C_1\right)\left(2\sigma_{xy} + C_2\right)}{\left(\mu_x^2 + \mu_y^2 + C_1\right)\left(\sigma_x^2 + \sigma_y^2 + C_2\right)} \tag{3.2}$$

where $C_1 = (K_1 * L)^2$, $C_2 = (K_2 * L)^2$, and L denotes the highest intensity level. The values of K_1 and K_2 are computed by philosophical study and are set to 0.01 and 0.03 respectively.

SSIM evaluates images on a structural domain rather than an error domain. Hence, for each template image, the score with respect to the target image becomes different. It ranges from 0 to 1 and a higher value indicates better matching. It has been observed from experiment that the recognition performance of all 211 classes is very high for

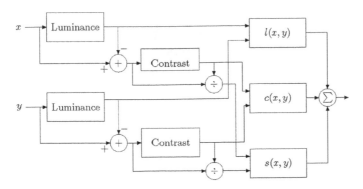

FIGURE 3.3 Schematic diagram of the SSIM system.

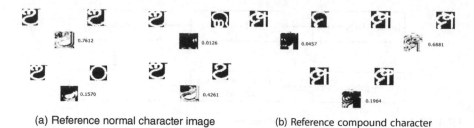

(a) Reference normal character image   (b) Reference compound character

FIGURE 3.4   Comparison of SSIM values of two template images.

the font that is similar to the template. But in the case of other fonts, the actual class can be found within the best 10–15 scores. For a fair evaluation, the best 20 characters that have high SSIM values are chosen and these are used in the second stage to refine the results. Figure 3.4 is an example of a working of SSIM on separating similar characters. Figure 3.4(a) shows how the target character �‍ gives different values for four alike templates that often creates confusion. An equivalent example for a compound character �\ is given in Figure 3.4(b).

3.2.3 PROJECTION PROFILE AND KENDALL RANK CORRELATION COEFFICIENT MATCHING

The projection profile analysis refers to the matching of the top, bottom, left, and right projection of the black pixel count in the corresponding character image. The projection profile of a sample character is depicted in Figure 3.5. With the change in font, the location of the outer ovals and the depression changes but the order in which they appear remains constant. The analysis of the projection can hence be considered

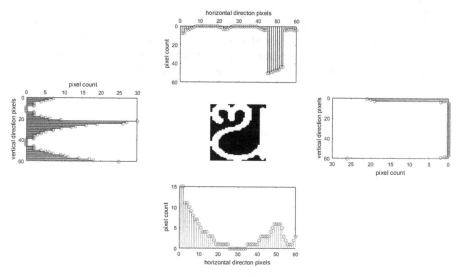

FIGURE 3.5   Projection profile of the character ঌ.

a rank-order-matching problem. For this, the Kendall rank correlation coefficient (KRCC) is used and Equation 3.3 describes KRCC values. Here, the target vector is formed by collecting the top, left, bottom, and right projection sequentially. The image is normalized to 60×60, hence the vector size becomes 1×240.

$$\tau = \frac{N_{cp} - N_{dp}}{n(n-1)/2} \tag{3.3}$$

where N_{cp} is the number of concordant pairs and N_{dp} is the number of discordant pairs. Let $\{X_1, X_2, \ldots, X_n\}$ be the observed/target vector and $\{Y_1, Y_2, \ldots, Y_n\}$ be the reference vector, then the joint random variable is $\{(X_1, Y_1), (X_2, Y_2), \ldots, (X_n, Y_n)\}$. A set of observations (X_i, Y_i) and (X_j, Y_j) is said to be concordant if both $X_i > X_j$ and $Y_i > Y_j$ or $X_i < X_j$ and $Y_i < Y_j$. Similarly a pair (X_i, Y_i) and (X_j, Y_j) is discordant if $X_i > X_j$ and $Y_i < Y_j$ or $X_i < X_j$ and $Y_i > Y_j$. If $X_i = X_j$ or $Y_i = Y_j$ then the pair is neither concordant nor discordant. The KRCC value varies from -1 to 1, where -1 represents being completely out of agreement, 1 represents being exactly the same, and zero represents the two vectors being independent. For matching characters, the value closest to 1 is preferred.

3.2.4 LOCAL FREQUENCY DESCRIPTOR

A local edge descriptor [23] has been derived from LFD [22]. Here we consider neighbor pixels with radius R. Let LCF = $\{f_1, \ldots, f_P\}$ be the set of neighborhood pixels that are located around the centroid. Hence LFD output can be expressed as:

$$LFD(n) = \sum_{k=1}^{P} f_k e^{\frac{-2\pi i(k-1)(n-1)}{P}}, \quad \text{for } (n = 1, \ldots, P) \tag{3.4}$$

where LFD(n) is a vector of $1 \times P$ corresponding to the frequency component of the neighborhood pixels in LCF. Among P components, LDF(2) contains the most edge information. Hence the local frequency descriptor gradient (LFDG) is calculated by setting $n = 2$:

$$LFDG = \sum_{k=1}^{P} f_k e^{\frac{-2\pi(k-1)}{P}} \tag{3.5}$$

LFDG is a complex value; therefore to use it in a neural network Equation 3.5 can be decomposed into real and imaginary parts as:

$$Re(LFDG) = \sum_{k=1}^{P} f_k \cos\left(\frac{-2\pi(k-1)}{P}\right) \tag{3.6}$$

$$Im(LFDG) = -\sum_{k=1}^{P} f_k \sin\left(\frac{-2\pi(k-1)}{P}\right) \tag{3.7}$$

Hence for each R the corresponding LFDG size is 1×2. For feature evaluation the character image is normalized to 60×60. LFDG is evaluated for $R = \{1, 2, ..., 30\}$ around its centroid, which makes the feature size to be 1×60.

3.2.5 GENERAL REGRESSION NEURAL NETWORK (GRNN)

The GRNN [45] was developed by Specht in 1991 and the main advantage of this network is that it needs no training. For a given training set (x_i, y_i), the estimation of unknown dependent variable y_{test} from independent variable x_{test} is given by:

$$y_{test} = \frac{\sum_{i=1}^{n} y_i \exp\left(-\frac{D_i^2}{2\sigma^2}\right)}{\sum_{i=1}^{n} \exp\left(-\frac{D_i^2}{2\sigma^2}\right)} \tag{3.8}$$

where σ is the smoothing factor (here chosen as $\sqrt{2}$), p is dimension X, n is the size of the training set, D_i is the distance of x_{test} to each respective x_i. Here Euclidean distance is used for calculating D_i, and is expressed as:

$$D_j^2 = \sqrt{\sum_{i=1}^{p} \left(x_{j,i} - x_{test,i}\right)^2} \tag{3.9}$$

Figure 3.6 shows the overall model of a GRNN network, which can handle the dynamical nature of the previous stage prediction. When multi-labels are predicted, the prediction is not always consistent, depending on the nature of the input. GRNN can compare and predict the output label in a single pass with high accuracy. This is based on the per-calculated samples from each class; hence it can handle noise. To fix the sample size we performed k-means on each class with the **k** value as 50. Fifty clustering centers are stored for each class to be used as the pattern input to the GRNN model.

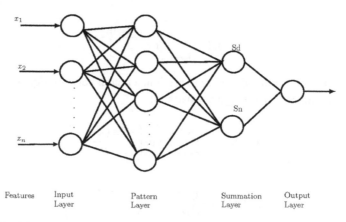

FIGURE 3.6 Block diagram for GRNN.

3.3 PROPOSED METHOD

The proposed prediction model consists of three stages. The first phase uses SSIM based template matching to reduce the number of possible prediction labels to 20. These probable character classes are further reduced to ten based on their geometrical features. In this stage, we used projection profile matching to match the similar characters geometrically. The final prediction of character is performed in the last stage by a machine learning approach. We employed LFD features and GRNN to predict the final class. The overall block diagram of the proposed three-stage hybrid model is depicted in Figure 3.7. Each stage of the model is described in Algorithms 3.1 to 3.3.

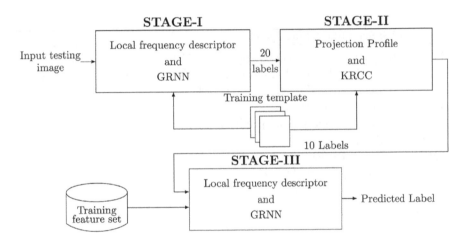

FIGURE 3.7   Block diagram for the proposed hybrid model.

Algorithm 3.1: Stage I detection algorithm

    Data: Template images $\{I_1, I_2, \ldots, I_m\}$ and Test character image (T)
           in preprocessed format
    Result: Best 20 probable characters
    for $I_k \leftarrow \{I_1, I_2, \ldots, I_m\}$ **do**
        Resize I_k and T to 60×60;
        $S_k \leftarrow SSIM(I_k, T)$;
    end
    Return best 20 character classes that have maximum S_k

Algorithm 3.2: Stage II detection algorithm

Data: 20 Template images $\{I_1, I_2, \ldots, I_{20}\}$ from Stage I and Test
   character image (T) in preprocessed format

Result: Best probable characters

$[Top_t, Left_t, Bottom_t, Right_t]$=ProjectionCount$(T)$;

for $I_k \leftarrow \{I_1, I_2, \ldots, I_{20}\}$ **do**

  $[Top_k, Left_k, Bottom_k, Right_k]$=ProjectionCount$(I_k)$;

  $\tau_k = $

   $KRCC([Top_t, Left_t, Bottom_t, Right_t], [Top_k, Left_k, Bottom_k, Right_k])$

end

Discard all option having 0 or -ve KRCC;

Return best at max 10 character classes that have maximum τ_k;

Algorithm 3.3: Stage III detection algorithm

Data: 10 Feature set for $\{(DI_1, Y_1), (DI_2, Y_2), \ldots, (DI_{10}, Y_{10})\}$ from
   Stage II and Test character image (T) in preprocessed format,
   where $DI_i = \{DI_i^1, DI_i^2, \ldots DI_i^{sz_i}\}$

Intialization: $\sigma = \sqrt{2}$

Result: Final character label prediction Y

$DI_T \leftarrow$ LFD(T);

for $i \leftarrow 1, 2, \ldots, 10$ **do**

  for *each sample j in* DI_i **do**

   $D_{i,j}^2 \leftarrow \sqrt{\sum_{k=1}^{p}(DI_i^{j,k} - DI_T^{j,k}}$;

   $\hat{Y}_{i,j} = onehotvector(i, 10)$;

  end

end

$$Y_p = \frac{\sum_{i=1}^{10}\sum_{j=1}^{sz_i} y_{i,j} exp(-\frac{D_{i,j}^2}{2\sigma^2})}{\sum_{i=1}^{10}\sum_{j=1}^{sz_i} exp(-\frac{D_{i,j}^2}{2\sigma^2})};$$

$x \leftarrow \arg_{max}(Y)$;

$Y \leftarrow Y_x$;

Return Y;

3.4 EXPERIMENTS

3.4.1 DATASET CREATION

From the earlier studies, we observed that most of the research on Odia character recognition is based on 47–51 basic characters. However, in Odia, each basic character can be combined with any other to produce a compound character and this makes the total number of classes to be the factorial of 51. But not all possible characters are used in Odia literature. From our study, we found a maximum of 432 characters that

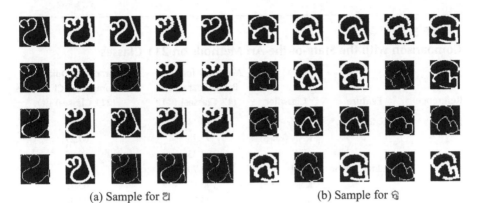

(a) Sample for ଥ   (b) Sample for ଡ଼

FIGURE 3.8   Sample images from the dataset.

are used, out of which 211 are used most of the time and the rest are used only rarely. Without recognition of these 211 characters, Odia OCR is not effective for real-world use. Moreover, the compound characters do not follow any specific rule of combination, hence making the recognition process more difficult. Since there are no datasets publicly available that cover any of the compound characters, we developed a dataset by collecting samples from each identified class. Our dataset covers font and size variation along with heavy print, thinning, and rotation errors. For each class, our dataset contains 250 samples which form a total sample size of 52,750. Some example images from the dataset are illustrated in Figure 3.8.

3.4.2 EXPERIMENTAL SETUP

For our experiment, we used a ten-fold cross-validation strategy to avoid the overfitting problem. Each of the methods and hybrid models is implemented using Matlab 2015b on a Windows machine. The following hyper-parameters were used for the proposed model. The first stage used SSIM for template matching, and the parameters for it have been kept fixed. In the second stage, KRCC ranking was used which does not require any hyper-parameters, while in the third stage the GRNN model was utilized, which is a one-pass learning approach and which required a smoothing factor to predict the class label of the unknown sample. Here, the smoothing factor was set as $\sqrt{2}$. We verified the effectiveness of the single-stage feature-based recognition models on both 47 classes and 211 classes of characters. It was observed that the existing methods achieved higher accuracy for 47 classes of characters; however, they earned comparatively lower accuracy while considering 211 classes. The comparative results are shown in Table 3.1. Hence, a multi-stage hybrid model is proposed to improve the recognition accuracy across both problems. We explored many hybrid models and found that the model that combines SSIM matching, projection profile ranking with KRCC, and GRNN with an LFD feature yields the best accuracy. The SSIM matching reduces the number of probable characters to 20 based on overall structural similarity. Then, the projection-profile based ranking method is employed to further reduce the probable characters to ten. Finally, LFD features and

TABLE 3.1

Comparison with the State-of-the-Art Methods on 211 Classes

Reference	Feature	Classifier	Accuracy for Odia Basic Characters (47 Classes) (%)	Accuracy for Odia Basic and Compound Characters (211 Classes) (%)
[5]	Zone feature	SVM	70.14	46.88
		KNN ($k = 5$)	91.78	67.27
		MLP	82.33	69.99
[24]	Quadrant feature	SVM	20.70	10.30
		KNN ($k = 5$)	96.50	65.25
		MLP	83.22	33.50
[38]	Zone mean feature (16 zone)	SVM	49.70	32.33
		KNN ($k = 5$)	98.63	61.62
		MLP	82.80	68.22
[35]	Directional feature	SVM	98.14	89.21
		KNN ($k = 5$)	97.91	88.83
		MLP	96.14	89.10
[40]	LU feature	SVM	50.21	18.57
		KNN($k = 5$)	81.34	10.71
		MLP	47.20	14.73
[26]	DCT	SVM	97.31	85.49
		KNN ($k = 5$)	97.71	87.81
		MLP	95.12	87.85
[48]	DCST + PCA	SVM	96.26	82.52
		KNN ($k = 5$)	96.70	89.50
		MLP	95.60	83.66
[13]	DOST + PCA	SVM	93.40	80.68
		KNN ($k = 5$)	95.05	86.90
		MLP	97.75	81.83
	Proposed Model		**97.95**	**90.60**

the GRNN classifier is employed to predict the final character label. The comparison results for different possible hybrid models are given in Table 3.2.

3.5 RESULTS AND DISCUSSION

From Table 3.1, it can be observed that the existing models obtain higher accuracy for 47 classes; however, they perform poorly while dealing with 211 character classes. A hybrid model helps in increasing the recognition accuracy using multiple stages. In this study, we proposed two different types of hybrid models based on the number of stages required: a two-stage hybrid model and a three-stage hybrid model. In the two-stage feature-based model, the recognition results of the first stage are further refined by using LFD and GRNN. In this case, different feature descriptors such as DCST and DOST along with an SVM classifier were taken into

TABLE 3.2
Recognition Performance of Different Possible Hybrid Models

Hybrid Model		Accuracy (%)
Two-stage hybrid classification	Stage I: DCST + PCA and SVM Stage II: LFD and GRNN	75.38
	Stage I: DCST + PCA and SVM Stage II: LFD and SVM	78.73
	Stage I: DOST + PCA and SVM Stage II: LFD and SVM	80.45
	Stage I: DOST + PCA and SVM Stage II: LFD and GRNN	83.64
Three-stage hybrid classification	Stage I: SSIM matching Stage II: Projection profile matching (KROCC) Stage III: LFD and GRNN	90.60
	Stage I: SSIM matching Stage II: Projection profile matching (KROCC) Stage III: LFD and SVM	80.80

consideration. However, these models achieve a reduced accuracy as compared to the single-stage models which are shown in Table 3.2. This is because stage I produces similar probability values for the characters with the same geometric structures. Hence, we designed a three-stage hybrid model to overcome the limitations of the two-stage model. Here, we used template matching and ranking methods to reduce the number of predicted character classes. Then, we employed SSIM matching and profile projection ranking to further reduce the highly improbable characters before predicting the final class using LED and GRNN/SVM. The results of the hybrid models are shown in Table 3.2. It is observed that the three-stage model with LED and GRNN provides the highest recognition performance (i.e., 90.6%) in comparison to the other hybrid models and state-of-the-art methods.

The proposed model is applied to form an OCR for Odia characters. Figures 3.9–3.14 show the results for a sample input and output at different stages of the proposed model.

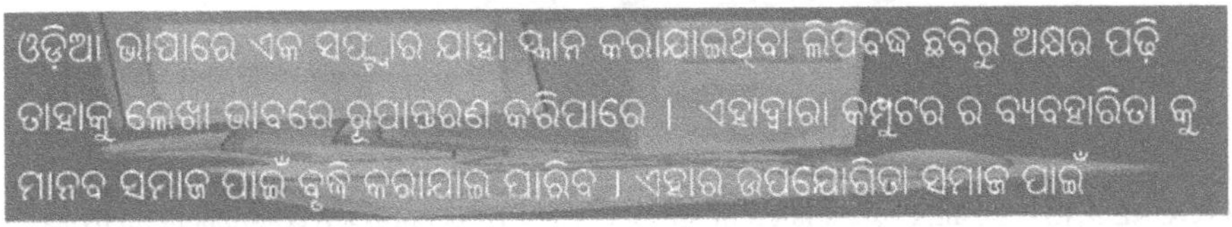

ଓଡ଼ିଆ ଭାଷାରେ ଏକ ସଫ୍ଟୱାର ଯାହା ସ୍କାନ କରାଯାଇଥିବା ଲିପିବଦ୍ଧ ଛବିରୁ ଅକ୍ଷର ପଢ଼ି
ତାହାକୁ ଲେଖା ଭାବରେ ରୂପାନ୍ତରଣ କରିପାରେ । ଏହାଦ୍ୱାରା କମ୍ପୁଟର ର ବ୍ୟବହାରିତା କୁ
ମାନବ ସମାଜ ପାଇଁ ବୃଦ୍ଧି କରାଯାଇ ପାରିବ । ଏହାର ଉପଯୋଗିତା ସମାଜ ପାଇଁ
(a)

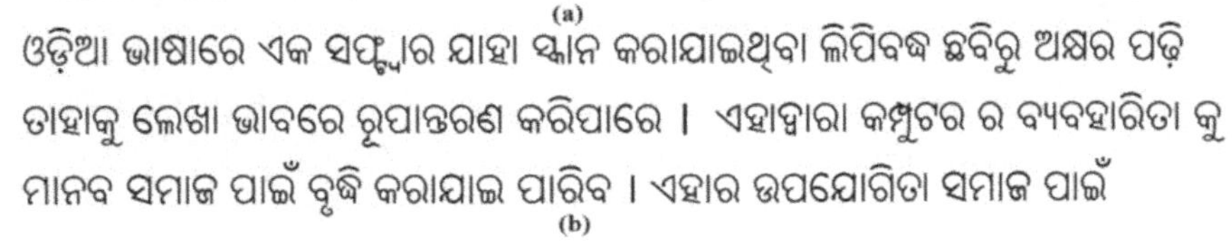

ଓଡ଼ିଆ ଭାଷାରେ ଏକ ସଫ୍ଟୱାର ଯାହା ସ୍କାନ କରାଯାଇଥିବା ଲିପିବଦ୍ଧ ଛବିରୁ ଅକ୍ଷର ପଢ଼ି
ତାହାକୁ ଲେଖା ଭାବରେ ରୂପାନ୍ତରଣ କରିପାରେ । ଏହାଦ୍ୱାରା କମ୍ପୁଟର ର ବ୍ୟବହାରିତା କୁ
ମାନବ ସମାଜ ପାଇଁ ବୃଦ୍ଧି କରାଯାଇ ପାରିବ । ଏହାର ଉପଯୋଗିତା ସମାଜ ପାଇଁ
(b)

FIGURE 3.9   Result of conversion of (a) original image and (b) binary image.

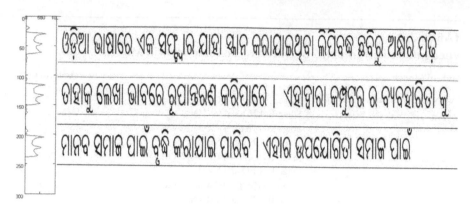

FIGURE 3.10   Result of line separation from binary image.

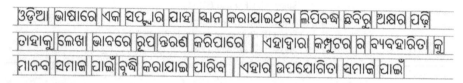

FIGURE 3.11   Result of word separation from each line of image.

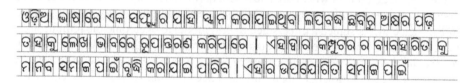

FIGURE 3.12   Result of separation of compound and character from each word.

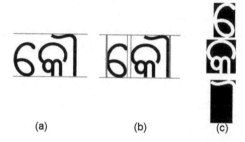

(a)   (b)   (c)

FIGURE 3.13   Result of separation of special symbol by connected component analysis.

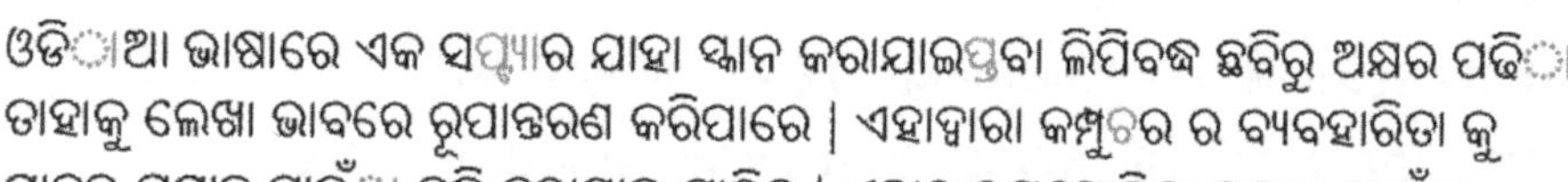

ଓଡ଼ିଆ ଭାଷାରେ ଏକ ସଂଖ୍ୟାର ଯାହା ସ୍ଥାନ କରାଯାଇଥିବା ଲିପିବଦ୍ଧ ଛବିରୁ ଅକ୍ଷର ପଢ଼ି
ତାହାକୁ ଲେଖା ଭାବରେ ରୂପାନ୍ତରଣ କରିପାରେ | ଏହାଦ୍ୱାରା କମ୍ପ୍ୟୁଟର ର ବ୍ୟବହାରିତା କୁ
ମାନବ ସମାଜ ପାଇଁ ବୃଦ୍ଧି କରାଯାଇ ପାରିବ | ଏହାର ଉପଯୋଗିତା ସମାଜ ପାଇଁ

FIGURE 3.14   Recognition output of the model.

Figure 3.9(a) shows the original input to OCR application with a background, whereas Figure 3.9(b) depicts the output of the binarization step. We used Tsallis entropy and a differential evolution optimization technique to separate the background from the foreground text. In the next stage, the lines are separated by line-wise histogram analysis. The output of the line separation is given in Figure 3.10. For each line, column-wise histogram analysis is performed, and a threshold distance for space is evaluated. Based on this, we separated words from each line and the output is illustrated in Figure 3.12. We used connected component analysis to separate characters from words and the corresponding output is shown in Figures 3.12 and 3.13. Each separated character is recognized by the classification model and the corresponding UNI-code is used to generate the text file. The final output of the application is given in Figure 3.14. The recognition errors in the text are highlighted in red.

In this study, we used 211 characters to validate the model in contrast to the previous models that considered only 47–51 characters. It was observed from the results how the existing models failed to perform on 211 characters and how our model helped to improve the recognition accuracy. The usability of the model was shown through an application example, which can be used for saving historical documents in digital form. It can also help to manage different printed documents by converting them into an editable text format.

3.6 CONCLUSION AND FUTURE SCOPE

In this chapter, we proposed a hybrid three-stage model for recognition of Odia compound and basic characters. The model first utilized the structural similarity index together with template matching to find 20 alike characters for a given testing sample. Subsequently, we applied a projection matching technique to obtain the most alike characters. Eventually, we predicted the actual class using LFD features and GRNN. We validated the proposed model using a dataset spanning 52,750 images from 211 classes. The proposed scheme obtained a higher accuracy of 90.6% when compared with the state-of-the-art methods and can hence be used to better recognize the historical and contemporary Odia documents. The earned accuracy is still far from achieving human-like precision, hence, in the future, further research needs to be carried out towards the improvement of model performance. The separation between the touched character and overlapping words could be made for accurate recognition of characters. Deep learning algorithms have recently achieved dramatic success in a variety of computer vision applications and, hence, the performance on the considered dataset could be tested using different contemporary deep learning algorithms.

REFERENCES

1. Alginahi, Y.: *Preprocessing Techniques in Character Recognition*. INTECH Open Access Publisher (2010).
2. Antani, S., Agnihotri, L.: Gujarati character recognition. In: *Proceedings of the Fifth International Conference on Document Analysis and Recognition*. pp. 418–421. Los Alamitos, CA: IEEE (1999).
3. Ashwin, T., Sastry, P.: A font and size-independent OCR system for printed Kannada documents using support vector machines. *Sadhana* 27(1), 35–58 (2002).

4. Basa, D., Meher, S.: Handwritten Odia character recognition. *Recent Advances in Microwave Tubes, Devices and Communication*, Jaipur, India (2011).
5. Bhowmik, T.K., Parui, S.K., Bhattacharya, U., Shaw, B.: An hmm based recognition scheme for handwritten Oriya numerals. In: *International Conference on Information Technology*. pp. 105–110. Bhubaneswar, India: IEEE (2006).
6. Chaudhuri, B., Pal, U.: A complete printed Bangla OCR system. *Pattern Recognition* 31(5), 531–549 (1998).
7. Chaudhuri, B., Pal, U., Mitra, M.: Automatic recognition of printed Oriya script. *Sadhana* 27(1), 23–34 (2002).
8. Cheriet, M., Kharma, N., Liu, C.L., Suen, C.: *Character Recognition Systems: A Guide for Students and Practitioners.* Hoboken, NJ: John Wiley & Sons (2007).
9. Chinnuswamy, P., Krishnamoorthy, S.G.: Recognition of handprinted Tamil characters. *Pattern Recognition* 12(3), 141–152 (1980).
10. Das, D., Nayak, D.R., Dash, R., Majhi, B.: An empirical evaluation of extreme learning machine: Application to handwritten character recognition. *Multimedia Tools and Applications* 78(14), 19495–19523 (2019).
11. Das, D., Nayak, D.R., Dash, R., Majhi, B., Zhang, Y.D.: H-wordnet: A holistic convolutional neural network approach for handwritten word recognition. *IET Image Processing* (March 2020), https://digital-library.theiet.org/content/journals/10.1049/iet-ipr.2019.1398.
12. Dash, K.S., Puhan, N., Panda, G.: A hybrid feature and discriminant classifier for high accuracy handwritten Odia numeral recognition. In *IEEE Region 10 Symposium*, Kuala Lumpur, Malaysia, pp. 531–535. IEEE (2014).
13. Dash, K.S., Puhan, N., Panda, G.: Non-redundant stockwell transform based feature extraction for handwritten digit recognition. In *International Conference on Signal Processing and Communications*, Bangkok, Thailand, pp. 1–4. IEEE (2014).
14. Dash, K.S., Puhan, N., Panda, G.: On extraction of features for handwritten Odia numeral recognition in transformed domain. In *Eighth International Conference on Advances in Pattern Recognition*, Kolkata, India, pp. 1–6. IEEE (2015).
15. Dash, K., Puhan, N., Panda, G.: BESAC: Binary external symmetry axis constellation for unconstrained handwritten character recognition. *Pattern Recognition Letters* 83, 413–422 (2016).
16. Dholakia, J., Yajnik, A., Negi, A.: Wavelet feature based confusion character sets for Gujarati script. In *International Conference on Conference on Computational Intelligence and Multimedia Applications*, vol. 2, pp. 366–370. IEEE (2007).
17. Garain, U., Chaudhuri, B.: Compound character recognition by run-number-based metric distance. In *Photonics West'98 Electronic Imaging*, San Jose, CA, pp. 90–97. International Society for Optics and Photonics (1998).
18. Khurshid, K., Siddiqi, I., Faure, C., Vincent, N.: Comparison of niblack inspired binarization methods for ancient documents. In *IS&T/SPIE Electronic Imaging*, San Jose, CA, pp. 72470U–72470U. International Society for Optics and Photonics (2009).
19. Kittler, J., Illingworth, J.: Minimum error thresholding. *Pattern Recognition* 19(1), 41–47 (1986).
20. Kumar, B., Kumar, N., Palai, C., Jena, P.K., Chattopadhyay, S.: Optical character recognition using ant miner algorithm: A case study on Oriya character recognition. *Int. J. Comput. Appl* 57(7), 33–41 (2012).
21. Lehal, G., Singh, C.: Feature extraction and classification for OCR of Gurmukhi script. *VIVEK-BOMBAY* 12(2), 2–12 (1999).
22. Maani, R., Kalra, S., Yang, Y.H.: Rotation invariant local frequency descriptors for texture classification. *IEEE Transactions on Image Processing* 22(6), 2409–2419 (2013).
23. Maani, R., Kalra, S., Yang, Y.H.: Robust volumetric texture classification of magnetic resonance images of the brain using local frequency descriptor. *IEEE Transactions on Image Processing* 23(10), 4625–4636 (2014).
24. Mahto, M.K., Kumari, A., Panigrahi, S.: Asystem for Oriya handwritten numeral recognization for Indian postal automation. *International Journal of Applied Science & Technology Research Excellence* 1(1), 17–23 (2011).

25. Meher, S., Basa, D.: An intelligent scanner with handwritten Odia character recognition capability. In *2011 Fifth International Conference on Sensing Technology (ICST)*, Palmerston North, New Zealand, pp. 53–59. IEEE (2011).
26. Mishra, T.K., Majhi, B., Panda, S.: A comparative analysis of image transformations for handwritten Odia numeral recognition. In *International Conference on Advances in Computing, Communications and Informatics*, Chennai, India, pp. 790–793. IEEE (2013).
27. Mishra, T.K., Majhi, B., Dash, R.: Shape descriptors-based generalised scheme for handwritten character recognition. *International Journal of Computational Vision and Robotics* 6(1–2), 168–179 (2016).
28. Mishra, T.K., Majhi, B., Sa, P.K., Panda, S.: Model based Odia numeral recognition using fuzzy aggregated features. *Frontiers of Computer Science* 8(6), 916–922 (2014).
29. Mohanty, S.: Pattern recognition in alphabets of Oriya language using Kohonen neural network. *International Journal of Pattern Recognition and Artificial Intelligence* 12(07), 1007–1015 (1998).
30. Mori, S., Suen, C.Y., Yamamoto, K.: Historical review of OCR research and development. In *Document Image Analysis*, pp. 244–273. IEEE Computer Society Press.
31. Mousa, M.A., Sayed, M.S., Abdalla, M.I.: An efficient algorithm for Arabic optical font recognition using scale-invariant detector. *International Journal on Document Analysis and Recognition (IJDAR)* 18(3), 263–270 (2015).
32. Padhi, D.: Novel hybrid approach for Odia handwritten character recognition system. *International Journal of Advanced Research in Computer Science and Software Engineering* 2(5), 150–157 (2012).
33. Padhi, D., Senapati, D.: Zone centroid distance and standard deviation based feature matrix for Odia handwritten character recognition. In *Proceedings of the International Conference on Frontiers of Intelligent Computing: Theory and Applications (FICTA)*, Odisha, India, pp. 649–658. Springer (2013)
34. Pal, U., Chaudhuri, B.: Indian script character recognition: A survey. *Pattern Recognition* 37(9), 1887–1899 (2004).
35. Pal, U., Sharma, N., Wakabayashi, T., Kimura, F.: Handwritten numeral recognition of six popular Indian scripts. In *Ninth International Conference on Document Analysis and Recognition (ICDAR 2007)*, Curitiba, Brazil, vol. 2, pp. 749–753 (September 2007).
36. Pal, U., Wakabayashi, T., Kimura, F.: A system for off-line Oriya handwritten character recognition using curvature feature. In *10th International Conference on Information Technology*, Beijing, China, pp. 227–229. IEEE (2007).
37. Pujari, P., Majhi, B.: A survey on Odia character recognition. *International Journal of Emerging Science and Engineering* 3(4), 15–25 (2015).
38. Roy, K., Pal, T., Pal, U., Kimura, F.: Oriya handwritten numeral recognition system. In *Eighth International Conference on Document Analysis and Recognition*, Seoul, Korea, pp. 770–774. IEEE (2005).
39. Sarangi, P. K., Ahmed, P.: Recognition of handwritten Odia numerals using artificial intelligence techniques. *International Journal of Computer Science* 2(02), 35–38 (2013).
40. Sarangi, P. K., Ahmed, P., Ravulakollu, K. K.: Naive Bayes classifier with LU factorization for recognition of handwritten Odia numerals. *Indian Journal of Science and Technology* 7(1), 35–38 (2014).
41. Sarangi, P. K., Sahoo, A. K., Ahmed, P.: Recognition of isolated handwritten Oriya numerals using Hopfield neural network. *International Journal of Computer Applications* 40(8), 36–42 (2012).
42. Sarkar, S., Das, S., Paul, S., Polley, S., Burman, R., Chaudhuri, S. S.: Multi-level image segmentation based on fuzzy-tsallis entropy and differential evolution. In *IEEE International Conference on Fuzzy Systems (FUZZ)*, Hyderabad, India, pp. 1–8. IEEE (2013).
43. Sauvola, J., Pietikiainen, M.: Adaptive document image binarization. *Pattern Recognition* 33(2), 225–236 (2000).

44. Sekita, I., Kurita, T., Otsu, N., Abdelmalek, N.: A thresholding method using the mixture of normal density functions. *International Symposium on Speech, Image Processing and Neural Networks* 1, 304–307 (1994).
45. Specht, D. F.: A general regression neural network. *IEEE Transactions on Neural Networks* 2(6), 568–576 (1991).
46. Sukhaswami, M., Seetharamulu, P., Pujari, A. K.: Recognition of Telugu characters using neural networks. *International Journal of Neural Systems* 6(03), 317–357 (1995).
47. Wang, Z., Bovik, A. C., Sheikh, H. R., Simoncelli, E. P.: Image quality assessment: From error visibility to structural similarity. *IEEE Transactions on Image Processing* 13(4), 600–612 (2004).
48. Mohapatra, R. K., Majhi, B., Jena, S. K.: Classification performance analysis of MNIST Dataset utilizing a multi-resolution technique. In *2015 International Conference on Computing, Communication and Security (ICCCS)*, Pamplemousses, pp. 1–5 (2015).

4 Development of Hybrid Computational Approaches for Landslide Susceptibility Mapping Using Remotely Sensed Data in East Sikkim, India

*Indrajit Chowdhuri, Paramita Roy,
Rabin Chakrabortty, Subodh Chandra Pal,
Biswajit Das, and Sadhan Malik*

CONTENTS

4.1 INTRODUCTION

A landslide is a natural disaster, which causes a loss of social as well as economic property. In recent decades human civilization has been increasing, but not in a scientific way and also in the process destroying the ecology of mountainous regions and enhancing the likelihood of landslides. A landslide is one type of slope instability process and also a mass movement (such as a rock fall, earth flow and debris fall) (Borgatti and Soldati 2010). It is also the outcome of the relationship between shear stress and the materials' shear strength on the slope (Yalcin 2007). In hilly and mountainous regions, landslide occurrences are very common and frequent in nature, but the time of their happening is not yet possible to determine (Glade and Crozier 2005). The presentation of landslide susceptibility mapping is an area where various rapid mass movement happens, and such an area is identified according to various geo-environmental factors (like slope, aspect, elevation, profile curvature, plan curvature, topographic wetness index (TWI), stream power index (SPI), geology, lithology, soil, distance to thrust, lineament density, land use and land cover (LULC), the normalized differences vegetation index (NDVI), road density, rainfall, and geomorphology) and present landslide area features. Fell et al. (2008) wrote that a susceptibility map shows the spatial distribution, thus the known and unknown landslide areas based on local terrain units. The preparation of a landslide susceptibility map of any region helps to forecast future landslide prone areas or vulnerable zones (Ahmed 2015; Pal et al. 2019). Traditionally, statistical approaches were used for large areas and recently trend machine learning models have also been widely used with their ensembles for occurrences and also for validation (Bui et al. 2016). The statistical methods include various approaches, such as logistic regression (LR), multivariate regression, a statistical index, an analytical hierarchy process (AHP), an evidential belief function (EBF), and certainty factors (Chakrabortty et al. 2018, 2020; Chowdhuri et al. 2020; Das and Pal 2019a, 2019b, 2020; Das et al. 2019a, 2019b; Pal et al. 2019). And therefore machine learning is an artificial intelligence branch which considers computer-based algorithms for making accurate predictions from the dominant data. Various previous literature reviews informed about used machine learning models for landslide prediction areas. These studies included an artificial neural network (ANN); a neural network model; decision trees (DTs); a support vector machine (SVM); boosted regression trees (BRTs); random forest (RF) biogeography-based optimization (BBO); eco-biogeography-based optimization (EBO); data mining; neuro-fuzzy and hybrid neuro-fuzzy, decision-making models; and big data analysis models (Das et al. 2015, 2018, 2019c, 2020a, 2020b, 2020c, 2020d; Dey et al. 2018, 2019; Rout et al. 2020).

Nowadays, hybrid machine-learning models are exploring the model in a scientific way. The hybrid models used in previous studies are ANN-fuzzy logic, stepwise weight assessment ratio analysis (SWARA), an adaptive neuro-fuzzy inference system (ANFIS), ANFIS with frequency ratio (FR), EBF-fuzzy logic, and our recent research work considers hybrid biogeography-based optimization ensembles with a main operator, and which are migrations and mutations. From a previous literature review we were assured that every approach has its own perspective statistical methods and has great influence on factor classes, while machine learning showed correlation between a landslide area and conditioning factors. Hybrid machine learning has two such types of implementation. For this reason its capability for identifying landslide-prone areas increases day by day. This hybrid model has another important capability: to evaluate landslide-related dependent and independent variables. The accuracy level and prediction power for identifying susceptible zones is higher than for the above two models (statistical and machine learning).

In hilly regions every year many news reports are published about landslides. If we check the global scale, then we can see almost 66 million people live in highly landslide-prone areas, where 17% of them are affected by this natural disaster (Sassa and Canuti 2008). Sikkim is a small hilly state in India facing in its southerly direction Eastern Himalaya, where most of the topography has developed by erosional process through a large number of perennial and non-perennial streams and springs (Pal and Chowdhuri 2019; Pal et al. 2019; Rawat et al. 2016; Tambe et al. 2012). Out of four districts, the east district is more populated and civilized, where 79.55% of the total population lives in an urban area. The trend line of rapid urbanization is increasing day by day and is concentrated around the capital of Gangtok. The years of major landslides in Sikkim are 1968, 1997, 2007, 2011, and 2018 (Kaur et al. 2019). Choubey (1995) has reported that about 36,000 people were killed by the 1968 landslide alone. In Sikkim in 1997, June 7 was a black day. Due to heavy rainfall on that day 5000 houses were washed away, 50 people were injured, and National Highway 31A was damaged. In 2007 landslides happened in the area of Taktse. And in 2011, Sikkim faced landslides due to an earthquake (Kaur et al. 2019). In the current study area East Sikkim is more susceptible to landslides because of high population pressure, rising urbanization, and/or effective environment conditioning factors; and this steep hilly region is very vulnerable to landslides (Chakraborty et al. 2011). Some landslide susceptibility works have been done in East Sikkim as well as Sikkim but the many landslide causative factors and ensemble machine learning methods have not applied to landslide susceptibility mapping. So the objective of this present research is to prepare a landslide susceptible map of East Sikkim. Hybrid BBO and its ensembles were used to show more vulnerable zone of landslides from which the Sikkim Government can take administrative steps to reduce rapid unscientific civilizations and also can manage increasing urban development in a scientific way.

4.2 STUDY MATERIALS AND METHODOLOGY

4.2.1 Area of Research Study

East Sikkim is one of the four administrative districts in Sikkim India. It is located in the south-eastern portion of the state. And this district is divided into the three subdivisions of Gangtok, Pakyong, and Rongli. The researched area covered 94 sq. km following the longitudinal and latitudinal extensions 88°27′ to 88°56′ E and 27°9′ to 27°25′ N respectively (Figure 4.1). Accordingly to the Indian Meteorological Department (IMD) East Sikkim belongs to the subtropical humid and temperate climate zone. As in the district of Sikkim, Teesta is the major river, and other major

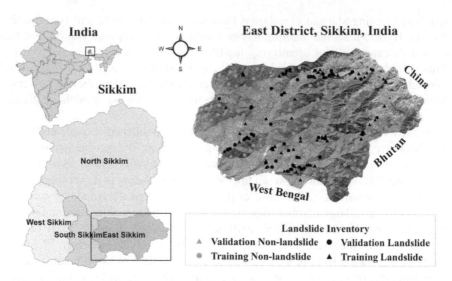

FIGURE 4.1   Location of the study area.

drainage systems are the Rangpo, Chhu, and Dik Chhu. The average rainfall is 3894 mm which mainly occurs during the period of May to September. Temperature varies in summer from 15°C to 20°C and in winter from 4°C to 10°C. Forest is the main land cover in this district and about 72.66 sq. km is its total area. Four types of forest coverage zones are found: mixed coniferous broad leaved forest, alpine coniferous, shrubs, and alpine meadow. Geomorphology consists of hilly, slope, and valley formations. The major soils are mountain meadow, brown red, yellow, and laterite. Predominant geological formations are recent alluvium Reyang, Gorubathan in the Daling group, migmatitic gneisses, augean gneiss, amphibolites in Darjeeling, and the Kanchenjunga group. This hilly area is mainly formed by gneiss, green schist, and amphibolites; they are all metamorphosed. The lineament of geological formation trends toward five directions: N–S, E–W, NE–SW, ENE–WSW, and NW–SE. Phylite, amphibolites, and schist were formed in the Precambrian era and alluvium along the river valley is from the Quaternary era.

4.2.2 Multi-colinearity Assessment (MCT)

By testing the colinearity among various affective factors of landslides we have obtained the independent variables for their occurring. The colinearity test shows the relationship between causative variables and strong prominent factors for landslides, and these factors can be separated from each other by their character. So if we don't do the colinearity test then the linear relationship of the variables reduces the accuracy of models. This process is controlled by two types of factors: the variance inflation factor (VIF) and the tolerance (TOL) index. The result of a multi-colinearity test establishes the linearity between causative factors when the coefficient value of TOL is ≤1 and the VIF factor is ≥10. Here, 18 factors are independent and differ in their characteristics, which helps to increase the accuracy of the models to establish a landslide susceptible map of East Sikkim.

4.2.3 Affecting Factors

To prepare a susceptibility map of the landslide area of East Sikkim, we selected 18 types of various geo-environment parameters which were sourced from ALOS PALSER DEM with 12.5 m and LANDSAT 8 at 30 m resolutions, meteorological data, a topographical map at 1:50,000, a thematic map, and locally available data from various Sikkim institutions working on landslide activity.

All considered factors were put into five sub-categories (Figure 4.2). Seven topographical factors (slope, elevation, aspect, plan curvature, profile curvature, TWI, and geomorphology) were sourced from a topographical map at 1:50,000 scale and from DEM at 12.5 m.

Three hydrological factors (Figure 4.3) (drainage density, rainfall, and SPI) are related to properties of hydrology; the information was collected from IMD, a topographical map, and a digital elevation model.

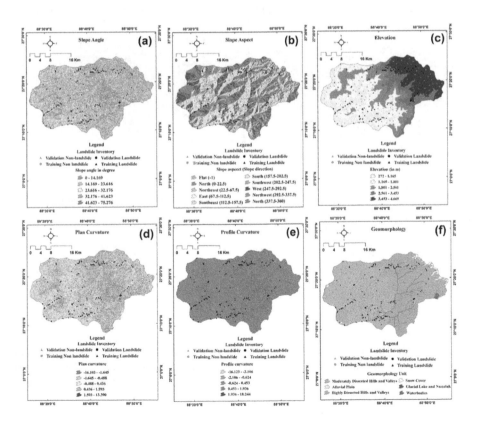

FIGURE 4.2   Topographical factors.

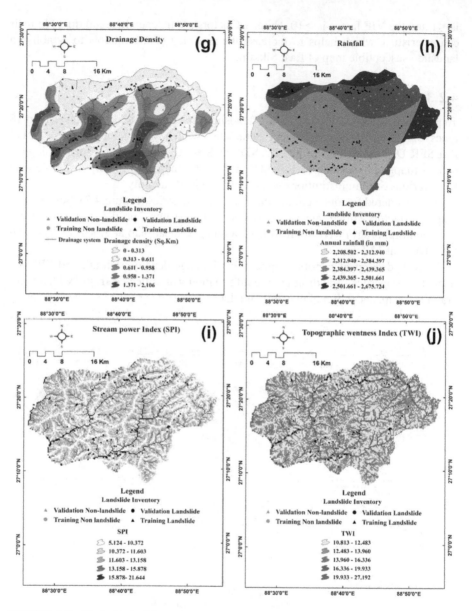

FIGURE 4.3 Hydrological factors.

Soil is another major element where its texture, compactation, and water holding capacity plays an important role in landslides. All soil related data are collected from the National Bureau of Soil Survey and Land Use Planning (NBSS&LUP) and the local survey office of Sikkim.

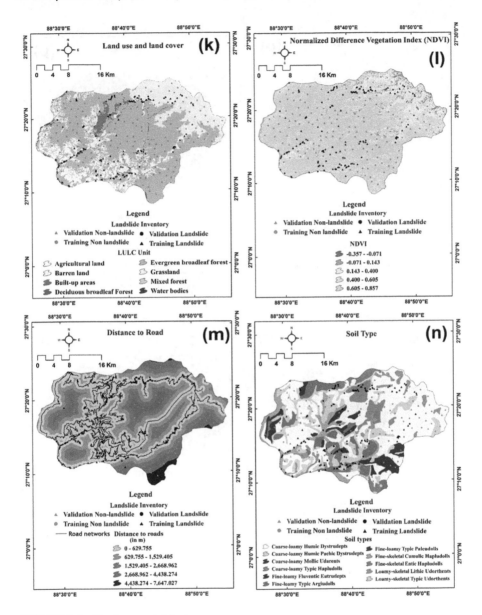

FIGURE 4.4 Environmental factors.

Environmental factors incorporating land use and land cover, NDVI, and road density were created from a topographical map and LANDSAT 8 images (Malik et al. 2019; Pal et al. 2019) (Figure 4.4). These factors may also be called anthropogenic influences because environments are affected by both nature as well as human activity.

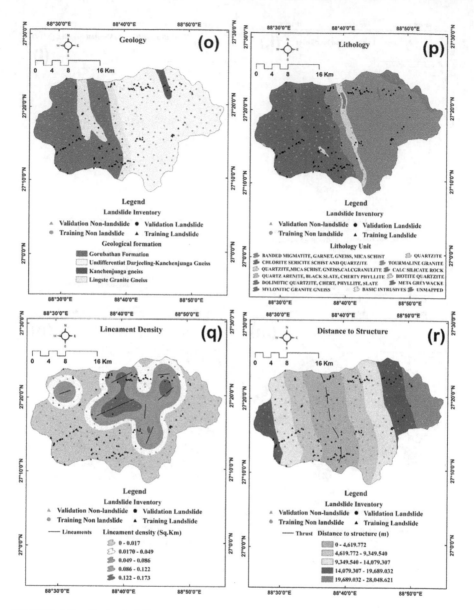

FIGURE 4.5   Geological factors.

Geology shows us the formation of the area and also the reason for the occurrence of slope-instability-related natural hazards. Distance to thrust, lineament density, and lithologies are all compactly developed main factors of geology (Figure 4.5). The Geological Survey of India (GSI) provided us with information about these factors to establish a geology map of East Sikkim.

4.2.4 Landslide Inventory Map (LIM)

Inventory maps as well as locational maps help to show the location of the researched study area. Here we show the landslide zone of East Sikkim by presenting an inventory map. Using predictive models and training data also shows us the spatial distribution of landslides and provides information for future landslide-vulnerable zones. We collected all data from the Sikkim Government and the GSI to prepare the landslide susceptible map and validated the results through a field survey and the Global Positioning System. Overall 163 landslide points are considered where 114 points (70%) are considered as the training dataset and 49 points (30%) are for validation.

4.2.5 Methodology

4.2.5.1 Hybrid Biogeography-Based Optimization

The hybrid BBO was evolved to form the fundamental structure of the BBO (Bhattacharya and Chattopadhyay 2010). The main way this was done was by devolving it by combining this algorithm with other ensembles. The BBO model is capable of estimating local exploration in a precise manner, but it is unable to estimate the level of exploration on a global scale (Simon 2008). So, the hybridization of BBO can be considered as a reliable prediction tool which is the best model that can balance global and local application (Savsani et al. 2014). For landslide susceptibility modeling, the landslide occurrence and non-landslide dataset were divided by the nominal data (such as 1 and 0). For Table 4.1, the frequency ratio (FR) value of each factor and sub-factor were needed for hybrid BBO modeling.

4.2.5.2 Hybridization with Differential Evolution

The diversified mutation scheme is very optimal, which makes differential evolution (DE) very realistic for determining large scale prediction when considering the maximum possible optimal solution. In a small region predictive capability is limited; so for this purpose mutation and migration can be used to complement each other. Beside this, there are several unique approaches that can be considered for the hybridization process.

4.2.5.2.1 The DE/BBO Algorithm

The integration of BBO with the DE algorithm (termed the DE/BBO algorithm) was initiated by Gong et al. (2010) (Equations 4.1–4.4). The basis of this optimization is the integration of hybrid BBO migration and DE mutation.

$$u_i \text{ of } D;$$

$$r_1, r_2, r_3 \in \left[1, N\right]$$

$$dr \in \left[1, D\right]$$

for $d = 1$ to D do

TABLE 4.1

Importance of Landslide Conditioning Factors

	Count	Area	Class (%)	Landslide Occurrence (%)	FR
Distance to Structure (m)					
0–4619.772	263,064	229.159	24.2125	25	1.03252
4619.772–9349.540	280,884	244.682	25.8527	32.8947	1.27239
9349.540–14079.307	271,031	236.099	24.9458	21.0526	0.84393
14079.307–19689.032	173,270	150.938	15.9479	13.1579	0.82506
19689.032–28048.621	98,298	85.6288	9.0474	7.89474	0.8726
Geology					
Gorubathan Fm.	379,386	330.488	34.9211	37.7193	1.08013
Undiff. Darjeeling – Kanchenjunga gneiss	565,641	492.738	52.0652	46.4912	0.89294
Kanchenjunga gneiss	15,047	13.1077	1.38502	1.31579	0.95001
Lingtse granite gneiss	126,473	110.172	11.6414	14.4737	1.2433
Geomorphology					
Moderately dissected hills and valleys	544,920	474.687	50.1958	54.386	1.08348
Alluvial plain	2212	1.9269	0.20376	0	0
Waterbody–river/waterbodies–other	26,176	22.8023	2.41123	2.63158	1.09139
Highly dissected hills and valleys	503,254	438.392	46.3577	42.5439	0.91773
Snow cover	9438	8.22157	0.86939	0	0
Glacial Lake and Nunatak	547	0.4765	0.05039	0.4386	8.70449
Lithology					
Banded migmatite, Garnet Bt gneiss, mica schist	535,088	466.123	49.2489	46.9298	0.95291
Quartzite	628	0.54706	0.0578	0.4386	7.58812
Tourmaline granite	72	0.06272	0.00663	0	0
Mylonitic granite gneiss	82,636	71.9854	7.60573	10.9649	1.44167
Basic intrusives	608	0.52964	0.05596	0	0
Chlorite sericite schist and quartzite	390,397	340.08	35.9317	36.4035	1.01313
Quartzite, mica schist, gneiss, calcgranulite	36,848	32.0988	3.39145	3.50877	1.03459
Calc silicate rock	5256	4.57858	0.48376	0	0
Unmapped	29,693	25.866	2.73291	0.4386	0.16049
Quartz arenite, black slate, cherty phyllite	569	0.49566	0.05237	0	0

Meta greywacke	1366	1.18994	0.12573	0.4386	3.48853
Dolimitic quartzite, chert, phyllite, slate	866	0.75438	0.07971	0	0
Biotite quartzite	2520	2.19521	0.23194	0.87719	3.78201
LUIC					
Agricultural land	182,066	158.6	16.5692	16.2281	0.97941
Barren land	125,131	109.003	11.3877	17.1053	1.50208
Built-up areas	20,594	17.9397	1.87419	2.63158	1.40412
Deciduous broadleaf forest	886	0.77181	0.08063	0	0
Evergreen broadleaf forest	673,446	586.648	61.288	55.7018	0.90885
Grassland	67,857	59.1112	6.17543	6.14035	0.99432
Mixed forest	7245	6.31122	0.65934	0.4386	0.6652
Water bodies	9322	8.12052	0.84836	1.75439	2.06797
Soil Texture					
Coarse-loamy mollic udarents	28,344	24.6909	2.6088	2.19298	0.84061
Loamy-skeletal lithic udorthents	48,896	42.594	4.50041	5.26316	1.16948
Loamy-skeletal typic udorthents	13,932	12.1364	1.28231	0.4386	0.34204
Fine-loamy fluventic eutrudepts	32,876	28.6387	3.02592	3.94737	1.30452
Coarse-loamy humic pachic Dystrudepts	176,575	153.817	16.2521	17.9825	1.10647
Coarse-loamy humic dystrudepts	528,617	460.486	48.6542	43.8596	0.90146
Fine-loamy typic paleudolls	20,264	17.6523	1.86511	1.31579	0.70548
Fine-loamy typic argiudolls	102,103	88.9434	9.39761	11.8421	1.26012
Fine-skeletal cumulic hapludolls	63,835	55.6076	5.87541	5.26316	0.89579
Fine-skeletal entic hapludolls	28,607	24.92	2.633	2.19298	0.83288
Coarse-loamy typic hapludolls	42,498	37.0206	3.91154	5.70175	1.45768
Aspect					
−1	20	0.01742	0.00184	0	0
0–22.5	48,897	42.5949	4.50022	4.38596	0.97461
22.5–67.5	76,454	66.6001	7.03642	6.57895	0.93499
67.5–12.5	97,792	85.188	9.00025	5.70175	0.63351
112.5–157.5	135,726	118.233	12.4915	14.4737	1.15868
157.5–202.5	151,908	132.329	13.9808	13.5965	0.97251
202.5–247.5	157,063	136.82	14.4552	19.2982	1.33503
247.5–292.5	168,057	146.397	15.4671	11.8421	0.76563
292.5–337.5	183,439	159.796	16.8827	17.5439	1.03916
337.5–360	67,191	58.531	6.1839	6.57895	1.06388

(Continued)

TABLE 4.1 (Continued)
Importance of Landslide Conditioning Factors

	Count	Area	Class (%)	Landslide Occurrence (%)	FR
Distance to Road (m)					
0–629.755	512,298	446.27	47.1522	64.9123	1.37666
629.755–1529.405	272,128	237.055	25.0468	17.1053	0.68293
1529.405–2668.962	188,440	164.153	17.3441	12.2807	0.70806
2668.962–4438.274	91,607	79.8001	8.43156	5.26316	0.62422
4438.274–7647.027	22,074	19.229	2.0317	0.4386	0.21588
Drainage Density (sq. km)					
0–0.313	325,531	283.575	29.962	26.3158	0.8783
0.313–0.611	271,389	236.411	24.9788	24.5614	0.98329
0.611–0.958	241,052	209.984	22.1866	21.9298	0.98843
0.958–1.371	142,226	123.895	13.0906	11.4035	0.87112
1.371–2.106	106,349	92.6421	9.78842	15.7895	1.61308
Elevation (m)					
272–1165	211,720	184.432	19.4856	21.4912	1.10293
1165–1801	293,733	255.875	27.0336	29.386	1.08702
1801–2561	223,326	194.542	20.5537	16.6667	0.81088
2561–3453	136,828	119.193	12.5929	14.4737	1.14935
3453–4669	220,940	192.464	20.3341	17.9825	0.88435
Lineament Density (sq. km)					
0–0.0170	573,776	499.824	52.8106	51.7544	0.98
0.0170–0.049	173,369	151.024	15.957	19.7368	1.23688
0.049–0.086	137,984	120.2	12.7001	11.4035	0.89791
0.086–0.122	132,661	115.563	12.2102	11.4035	0.93393
0.122–0.173	68,757	59.8952	6.32843	5.70175	0.90097
NDVI					
−0.357 to −0.071	35,069	3.5069	0.37053	0.4386	1.18369
−0.071 to 0.143	385,102	38.5102	4.06894	8.33333	2.04804
0.143–0.400	2,021,402	202.14	21.3579	27.6316	1.29374
0.400–0.605	2,668,625	266.863	28.1963	29.8246	1.05775
0.605–0.857	4,354,238	435.424	46.0063	33.7719	0.73407

Plan Curvature					
−16.103 to −1.645	56,047	48.8233	5.15827	5.70175	1.10536
−1.645 to −0.488	243,982	212.536	22.4548	16.6667	0.74223
−0.488 to 0.436	457275	398.339	42.0852	44.2982	1.05259
0.436–1.593	263,892	229.88	24.2872	27.193	1.11964
1.593–13.390	65,351	56.9282	6.01456	6.14035	1.02091
Profile Curvature					
−16.123 to −2.106	54,621	47.5811	5.02703	7.01754	1.39596
−2.106 to −0.624	231,429	201.601	21.2995	18.4211	0.86486
−0.624 to 0.453	443,031	385.93	40.7742	41.2281	1.01113
0.453–1.936	291,228	253.693	26.8031	27.193	1.01455
1.936–18.244	66,238	57.7008	6.09619	6.14035	1.00724
Rainfall (mm)					
2208.502–2312.940	77,742	67.7221	7.15541	7.01754	0.98073
2312.940–2384.397	216,875	188.923	19.9613	24.1228	1.20848
2384.397–2439.365	414,894	361.42	38.1871	33.7719	0.88438
2439.365–2501.661	238,173	207.476	21.9216	25.8772	1.18044
2501.661–2675.724	138,863	120.965	12.781	9.21053	0.72064
Slope (in Degrees)					
0–14.169	150,004	130.671	13.8056	10.9649	0.79424
14.169–23.616	288,534	251.346	26.5551	19.2982	0.72672
23.616–32.176	302,336	263.369	27.8254	27.193	0.97727
32.176–41.623	235,651	205.279	21.6881	29.386	1.35494
41.623–75.276	110,022	95.8417	10.1258	13.1579	1.29944
SPI					
5.124–10.372	5978	337.992	35.9969	35.5263	0.98693
10.372–11.603	5878	332.338	35.3947	33.7719	0.95415
11.603–13.158	3319	187.654	19.9855	17.5439	0.87783
13.158–15.878	1213	68.5821	7.30415	8.33333	1.1409
15.878–21.644	354	20.0149	2.13163	4.82456	2.26332
TWI					
10.813–12.483	7582	428.68	45.6554	41.2281	0.90303
12.483–13.960	5981	338.161	36.0149	34.6491	0.96208
13.960–16.336	2117	119.694	12.7476	12.2807	0.96337
16.336–19.933	816	46.136	4.91359	6.14035	1.24967
19.933–27.192	246	13.9087	1.4813	5.70175	3.84915

$$\text{if } rand < \lambda_i \text{ then}$$

$$rand < c_r \text{ or } d = d_r \tag{4.1}$$

$$\text{DE mutation} = (d) = hr1(d) + F.\left(Hr2(d) - Hr3(d)\right); \tag{4.2}$$

$$\text{BBO migration} = H_j \propto u_j; \tag{4.3}$$

$$u_i(d) = H_j(d); \tag{4.4}$$

where N is the population, D is the problem dimension, c is the crossover likelihood, and F is the scaling factor of mutation.

DE/BBO can replace the original mutation by considering the hybrid migration. For the purpose of calculating the migrant rate operator, the DE/BBO differs from the original BBO. The DE/BBO can be very optimistic in comparison to DE and BBO.

4.2.5.2.2 Local-DE/BBO

The improvement of BBO can be made with the help of local topologies. The DE/BBO with local topologies can be considered as a reliable predictor of hybrid BBO, which creates three categories of DE/BBO: RingDB, SquareDB, and RandDB. RingDB encompasses ring topology, whereas SquareDB and RandDB are related to square topology and random topology (Zheng et al., 2019). This triple local algorithm is more optimistic and is an improvement on DE/BBO. It is much more realistic than DE on the dimensional function:

$$\text{DE mutation} = u_i(d) = Hr_2(d) + F \ldots \left(Hr_2(d) - Hr_3(d)\right) \tag{4.5}$$

4.2.5.2.3 Self-Adaptive DE/BBO

Self-adaptive DE/BBO is an upgraded form of DE, and the main function of this model is in dealing with multiple mutation and choosing the most suitable one. The suitable mutation is very optimistic in functional analysis. The appropriate mutation combinedly works with the relevant parameters. A general approach for obtaining the information from initial values is:

$$V = V + \frac{t}{t_{\max}}\left(V_{\max} - V_{\min}\right) \tag{4.6}$$

where t is the current, max is the maximum, and min is the minimum generation number of the algorithm.

The obtained probability of the mutation and migration are estimated by:

$$P_k(G) = \frac{S_k(G)}{\displaystyle\sum_{k \in S_{DE}} S_k(G)} \tag{4.7}$$

$$P_k(G) = \frac{S_k(G)}{\sum_{k' \in S_{DE}} S_{k'}(G)} \tag{4.8}$$

The general function was replaced by the immigration rate and their associated cross-cover operations are:

$$f(x) = \begin{cases} Vij, (rand \le \lambda_i V(j = r_i) \\ Xij, \text{otherwise} \end{cases} \tag{4.9}$$

For increasing the consistency of the parameters, the Gaussian distribution is used:

$$F = norm\left(F_\mu, F_\sigma\right) \tag{4.10}$$

In HSDB, we set F_μ to 0.5 and F_σ to 0.3, and limit the value F in the range $[0.05, 0.95]$:

$$F = \begin{cases} 0.05, & F < 0.05 \\ 0.05, & F > 0.95 \\ 0.05, & 0.05 \le F \le 0.95 \end{cases} \tag{4.11}$$

4.2.6 VALIDATION OF MODELS

For the validation of the training dataset the AUC values with the receiver operating characteristic (ROC) curve was considered to obtain an authentic result. The AUC values state the model which is perfect for this area. Here, the ROC curve is a technique whose value is able to predict the best model. The scientific calculation is:

$$S_{AUC} = \sum_{k=1}^{n} (X_{k+1} - X_k)\left(S_k + 1 - S_{k+1} - \frac{S_k}{2}\right) \tag{4.12}$$

where S_{AUC} is considered an ROC, X_K is 1 specificity, and S_K is the sensitivity of ROC, which considers primary information about recent landslides. The performances of the classifiers identified the optimal model which is generated by different cutting threshed values.

4.2.7 SHORTLY STRUCTURED METHODOLOGY

This study consists of four main stages which are described and summarized here: data selection, data preparation, geo-conditioning factors, and a colinearity test using VIF and TOL for independent parameters selection. A landslide susceptibility map uses a hybrid BBO and its ensembles. The result is validated through AUC values of ROC. The prescribed methodology framework is shown in Figure 4.6.

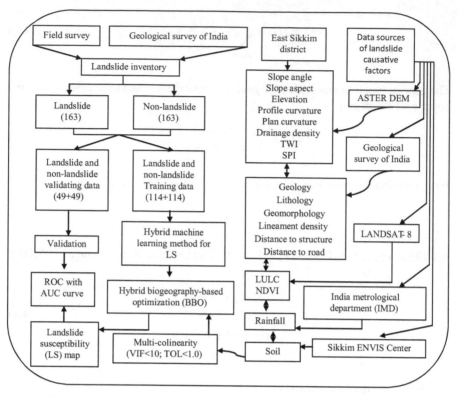

FIGURE 4.6   Methodology flow chart.

4.3 RESULTS AND DISCUSSION

4.3.1 IMPORTANCE OF THE CONDITIONING FACTORS ON THE OCCURRENCES OF LANDSLIDES

After the initial analysis the importance of all causative factors and their associated sub-factors was assessed with the help of frequency ratio. In the case of distance to structure, the high weightages are confirmed to lie between 0 to 4619.772 (1.033) and 4619.772 to 9349.540 (1.272). For the geology, high importance was associated with the Lingste granite gneiss (1.243) and the Gorubathan formation (1.080). In the case of geomorphology, high importance was associated with Glacial Lake and Nunatak (8.704), moderately dissected hills and valleys (1.083), and waterbody–river/waterbodies–other (1.091). In lithology, high importance was associated with quartzite (7.588), mylonitic granite gneiss (1.442), meta greywacke (3.489), and biotite quartzite (3.782). In LULC, high importance was associated with: barren land (1.502), built-up areas (1.404), and water bodies (2.068). In soil texture, high importance was associated with: loamy-skeletal lithic udorthents (1.169), fine-loamy fluventic eutrudepts (1.305), coarse-loamy humic pachic dystrudepts (1.106), fine-loamy typic argiudolls (1.260), and coarse-loamy typic hapludolls (1.458). In slope aspect, high importance was associated with the south-east (1.159), south-west (1.335), north-east (1.039), and north (1.064). In the case of the distance to road, the nearest distance such as 0–629.755 m (1.377) is favorable for the occurrence of landslides. Very high drainage density such as 1.371–2.106 (1.613) is favorable for the occurrence of landslides. In elevation, high importance was associated with 272–1165 m (1.103), 1165–1801 m (1.087), and 2561–3453 m (1.149). In lineament density, high importance was associated with 0.0170–0.049 sq. km (1.237) which is very optimistic regarding the occurrence of landslides. In NDVI, high importance was associated with −0.357 to −0.071 (1.184), −0.071 to 0.143 (2.048), and 0.143 to 0.400 (1.294). Here the lower values of NDVI indicate higher occurrences of landslides. In plan curvature, high importance was associated with −16.103 to −1.645 (1.105), −0.488 to 0.436 (1.053), 0.436 to 1.593 (1.120), and 1.593 to 13.390 (1.021). In profile curvature, high importance was associated with −16.123 to −2.106 (1.396) which is very optimistic regarding the occurrence of landslides in this region.

In the case of rainfall, high importance was associated with 2312.940–2384.397 mm (1.208) and 2439.365–2501.661 (1.180). In the case of slope angle, high weightage was associated with 32.176–41.623 (1.355) and 41.623–75.276 (1.299). In SPI, high importance was associated with 13.158–15.878 (1.141) and 15.878–21.644 (2.263); here the positive relationship is associated between the higher SPI and occurrences of landslides. In the case of TWI, high importance was associated with 16.336–19.933 (1.250) and 19.933–27.192 (3.849); here the positive relationship is associated between the higher TWI and occurrences of landslides (Table 4.1).

4.3.2 APPLICATION OF HYBRID BIOGEOGRAPHY-BASED OPTIMIZATION FOR LANDSLIDE SUSCEPTIBILITY ASSESSMENT

Different causal factors have been considered for estimation of the landslide susceptible areas using the hybrid BBO model on a GIS platform. This was constructed on the basis of all causative factors and landslide inventory points. The landslide point is randomly split into a 70:30 ratio as training and validation datasets. The very high (0.729–1.00) landslide susceptible areas are found mainly in the middle portion of this region and its areal coverage is 91.70 sq. km (9.69%). High (0.517–0.729) landslide susceptible areas are found mainly in the northern, middle, and eastern portions of this region and its areal coverage is 128.65 sq. km (13.59%) (Figure 4.7). Moderate (0.309–0.517) landslide susceptible areas are found mainly in the northern, middle, western, and eastern portions of this region and its areal coverage is 143.39 sq. km (15.15%). Low (0.113–0.309) landslide susceptible areas are found mainly in the northern, middle, western, and eastern portions of this region and its areal coverage is 158.95 sq. km (16.79%). Very low (0.00–0.113) landslide susceptible areas are found in most portions of this region and its areal coverage is 423.81 sq. km (44.78%) (Figure 4.8).

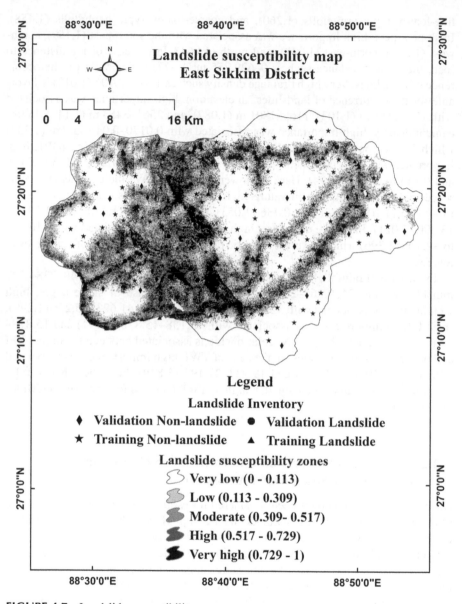

FIGURE 4.7   Landslide susceptibility map.

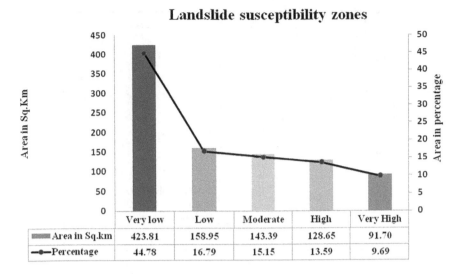

Landslide susceptibility zones

	Very low	Low	Moderate	High	Very High
Area in Sq.km	423.81	158.95	143.39	128.65	91.70
Percentage	44.78	16.79	15.15	13.59	9.69

FIGURE 4.8   Landslide susceptibility areas.

4.4 CONCLUSION

This research work presents a landslide susceptibility map to predict present and future hazard locations. Nowadays researchers are trying to investigate landslide susceptible zones using multiple models. Here the hybrid BBO machine learning model has been used to obtain a susceptibility map through the information about known and unknown landslide points. The model presented a spatial database by considering different geo-environment factors. From the AUC values of ROC it is established that hybrid BBO for the success rate and prediction rate are 0.9237 and 0.8979, respectively (Figure 4.9). All primary and secondary data were used to prepare the susceptibility map with the help of a hybrid BBO with three ensemble models. From that it is seen that there is a high tendency to occurrences of landslides and that the causative factors are very strong in creating such types of severe hazards. This hybrid machine learning is now used for many susceptibility and potentiality analyses. The hybrid BBO is also used for other susceptibilities like flood, soil erosion and gully erosion, and potentiality analysis for groundwater, minerals, and so on. The major limitations of this study is that the hybrid BBO model cannot give the major factors for landslide occurrences and, due to the physical barrier of east Sikkim, many inaccessible landslide locations have not been considered. Landslide-related information and landslide susceptibility mapping will help the government as well as researchers to take steps in hazard management and stabilizing natural conditions.

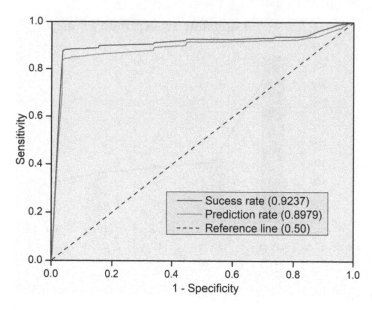

FIGURE 4.9 Validation.

REFERENCES

Ahmed, B. "Landslide susceptibility mapping using multi-criteria evaluation techniques in Chittagong Metropolitan Area, Bangladesh." *Landslides* 12, no. 6 (2015): 1077–1095.

Bhattacharya, A. and P. K. Chattopadhyay. Hybrid differential evolution with biogeography-based optimization for solution of economic load dispatch *IEEE Transactions on Power Systems* 25, no. 4 (2010): 1955–1964.

Borgatti, L. and M. Soldati. "Landslides as a geomorphological proxy for climate change: A record from the Dolomites (northern Italy)." *Geomorphology* 120, no. 1–2 (2010): 56–64.

Bui, D. T., T. A. Tuan, H. Klempe, B. Pradhan, and I. Revhaug. Spatial prediction models for shallow landslide hazards: A comparative assessment of the efficacy of support vector machines, artificial neural networks, kernel logistic regression, and logistic model tree *Landslides* 13, no. 2 (2016): 361–378.

Chakraborty, I., S. Ghosh, D. Bhattacharya, and A. Bora. "Earthquake induced landslides in the Sikkim-Darjeeling Himalayas—An aftermath of the 18th September 2011 Sikkim earthquake." Earth Quake Report, The best independent Earthquake reporting site in the world (2011): 1–8.

Chakrabortty, R., S. C. Pal, I. Chowdhuri, S. Malik, and B. Das. "Assessing the importance of static and dynamic causative factors on erosion potentiality using SWAT, EBF with uncertainty and plausibility, logistic regression and novel ensemble model in a subtropical environment." *Journal of the Indian Society of Remote Sensing* 48 (2020): 765–789.

Chakrabortty, R., S. C. Pal, S. Malik, and B. Das. "Modeling and mapping of groundwater potentiality zones using AHP and GIS technique: A case study of Raniganj Block, Paschim Bardhaman, West Bengal." *Modeling Earth Systems and Environment* 4, no. 3 (2018): 1085–1110.

Choubey, V. D. "Landslide hazards and their mitigation in the Himalayan region." In *International Symposium on Landslides*, Japan, pp. 1849–1868, 1995.

Chowdhuri, I., S. C. Pal, and R. Chakrabortty. "Flood susceptibility mapping by ensemble evidential belief function and binomial logistic regression model on river basin of eastern India." *Advances in Space Research* 65, no. 5 (2020): 1466–1489.

Das, B. and S. C. Pal. "Assessment of groundwater recharge and its potential zone identification in groundwater-stressed Goghat-I block of Hugli District, West Bengal, India." *Environment, Development and Sustainability* (2019a): 1–19.

Das, B. and S. C. Pal. "Assessment of groundwater vulnerability to over-exploitation using MCDA, AHP, fuzzy logic and novel ensemble models: A case study of Goghat-I and II blocks of West Bengal, India." *Environmental Earth Sciences* 79, no. 5 (2020): 1–16

Das, B. and S. C. Pal. "Combination of GIS and fuzzy-AHP for delineating groundwater recharge potential zones in the critical Goghat-II block of West Bengal, India." *HydroResearch* 2 (2019b): 21–30.

Das, B., S. C. Pal, S. Malik, and R. Chakrabortty. "Modeling groundwater potential zones of Puruliya district, West Bengal, India using remote sensing and GIS techniques." *Geology, Ecology, and Landscapes* 3, no. 3 (2019a): 223–237.

Das, B., S. C. Pal, S. Malik, and R. Chakrabortty. "Living with floods through geospatial approach: A case study of Arambag CD Block of Hugli District, West Bengal, India." *SN Applied Sciences* 1, no. 4 (2019b): 329.

Das, H., A. K. Jena, J. Nayak, B. Naik, and H. S. Behera. "A novel PSO based back propagation learning-MLP (PSO-BP-MLP) for classification." In L.C. Jain, H.S. Behera, J.K. Mandal, and D.P. Mohapatra (Eds.), *Computational Intelligence in Data Mining*, Volume 2, pp. 461–471. Springer, New Delhi, 2015.

Das, H., B. Naik, and H. S. Behera. "A hybrid neuro-fuzzy and feature reduction model for classification *Advances in Fuzzy Systems* 2020 (2020a): 1–15.

Das, H., B. Naik, and H. S. Behera. "An experimental analysis of machine learning classification algorithms on biomedical data." In *Proceedings of the 2nd International Conference on Communication, Devices and Computing*, pp. 525–539. Springer, Singapore, 2020b.

Das, H., B. Naik, and H. S. Behera. "Classification of diabetes mellitus disease (DMD): A data mining (DM) approach." In *Progress in Computing, Analytics and Networking*, pp. 539–549. Springer, Singapore, 2018.

Das, H., B. Naik, and H. S. Behera. "Medical disease analysis using neuro-fuzzy with feature extraction model for classification." *Informatics in Medicine Unlocked* 18 (2020c): 100288.

Das, H., B. Naik, H. S. Behera, S. Jaiswal, P. Mahato, and M. Rout. "Biomedical data analysis using neuro-fuzzy model with post-feature reduction." *Journal of King Saud University-Computer and Information Sciences* (2020d).

Das, H., N. Dey, and V. E. Balas, eds. *Real-Time Data Analytics for Large Scale Sensor Data*. Academic Press, London, UK, 2019c.

Dey, N., A. S. Ashour, H. Kalia, R. T. Goswami, and H. Das, eds. *Histopathological Image Analysis in Medical Decision Making*. IGI Global, Hershey, PA, 2018.

Dey, N., H. Das, B. Naik, and H. S. Behera, eds. *Big Data Analytics for Intelligent Healthcare Management*. Academic Press, Boston, MA, 2019.

Fell, R., J. Corominas, C. Bonnard, L. Cascini, E. Leroi, and W. Z. Savage. "Guidelines for landslide susceptibility, hazard and risk zoning for land-use planning." *Engineering Geology* 102, no. 3–4 (2008): 99–111.

Glade, T. and M. J. Crozier. "The nature of landslide hazard impact." In *Landslide Hazard and Risk*. Wiley, Chichester, 2005, pp. 43–74.

Gong, W., Z. Cai, and C. X. Ling. "DE/BBO: A hybrid differential evolution with biogeography-based optimization for global numerical optimization." *Soft Computing* 15, no. 4 (2010): 645–665.

Kaur, H., S. Gupta, S. Parkash, R. Thapa, A. Gupta, and G. C. Khanal. "Evaluation of landslide susceptibility in a hill city of Sikkim Himalaya with the perspective of hybrid modelling techniques." *Annals of GIS* 25, no. 2 (2019): 113–132.

Malik, S., S. Chandra Pal, B. Das, and R. Chakrabortty. "Assessment of vegetation status of Sali River basin, a tributary of Damodar River in Bankura District, West Bengal, using satellite data." *Environment, Development and Sustainability* (2019): 1–35.

Pal, S. C. and I. Chowdhuri. GIS-based spatial prediction of landslide susceptibility using frequency ratio model of Lachung River basin, North Sikkim, India. *SN Applied Sciences* 1, no. 5 (2019): 416.

Pal, S. C., B. Das, and S. Malik. "Potential landslide vulnerability zonation using integrated analytic hierarchy process and GIS technique of upper rangit catchment area, West Sikkim, India." *Journal of the Indian Society of Remote Sensing* 47, no. 10 (2019): 1643–1655.

Pal, S. C., R. Chakrabortty, S. Malik, and B. Das. "Application of forest canopy density model for forest cover mapping using LISS-IV satellite data: A case study of Sali watershed, West Bengal." *Modeling Earth Systems and Environment* 4, no. 2 (2018): 853–865.

Rawat, M. S., V. Joshi, D. P. Uniyal, and B. S. Rawat. Investigation of hill slope stability and mitigation measures in Sikkim Himalaya *International Journal of Landslide and Environment* 3, no. 1–3 (2016): 8–15.

Rout, M., A. K. Jena, J. K. Rout, and H. Das. "Teaching–learning optimization based cascaded low-complexity neural network model for exchange rates forecasting. In *Smart Intelligent Computing and Applications*, pp. 635–645. Springer, Singapore, 2020.

Sassa, K. and P. Canuti, eds. *Landslides-Disaster Risk Reduction*. Springer Science & Business Media, Berlin, 2008.

Savsani, P., R. L. Jhala, and V. Savsani. "Effect of hybridizing biogeography-based optimization (BBO) technique with artificial immune algorithm (AIA) and ant colony optimization (ACO)." *Applied Soft Computing* 21 (2014): 542–553.

Simon, D. "Biogeography-based optimization." *IEEE Transactions on Evolutionary Computation 12*, no. 6 (2008): 702–713.

Tambe, S., G. Kharel, M. L. Arrawatia, H. Kulkarni, K. Mahamuni, and Anil K. Ganeriwala. "Reviving dying springs: Climate change adaptation experiments from the Sikkim Himalaya." *Mountain Research and Development* 32, no. 1 (2012): 62–73.

Yalcin, A. "The effects of clay on landslides: A case study." *Applied Clay Science* 38, no. 1-2 (2007): 77–85.

Zheng, Y., X. Lu, M. Zhang, and S. Chen. *Biogeography-Based Optimization: Algorithms and Applications*. Springer Nature, Singapore, 2019.

5 Domain-Specific Journal Recommendation Using a Feed Forward Neural Network

S. Nickolas and K. Shobha

CONTENTS

5.1 INTRODUCTION

Information Technology (IT) has laid a strong foundation and has achieved greater heights in terms of electronic literature, leading to vast amounts of data (Zhao, Wu, and Liu 2016), which in turn leads to information overload problems (Drachsler, Hummel, and Koper 2008). With the enormous amount of data and growing competition in the research environment, scientific web archives have become more and more significant to different users (Meschenmoser, Meuschke, Hotz, and Gipp 2016). But, due to the enormous amount of data in scientific web archives, searching for a most relevant research article has become challenging and time-consuming, despite significant advances in digital libraries and information retrieval systems.

Research article recommendation demands a method that suggests significant articles to researchers via discovering their interests and preferences (Basu, Hirsh, Cohen, and Nevill-Manning 2001). By dynamically suggesting impressive materials, research article recommendations can save much time and effort for researchers (Pan and Li 2010).

Researchers are highly motivated to make their research work accessible to proliferate their research work, to increase their reputation, and to progress in their profession. Research and funding organizations often bank on scientific web archives, for example, to know the number of published papers and their citation counts for a researcher, and to appoint, promote, and to make funding decisions.

Domain-centric research article recommendations can help improve the quality and efficiency of the recommendation process by suggesting published articles similar to researchers' interests. For this drive, we methodically queried the web archives, scraped the returned links, and inspected different features that contribute to the recommendation of publications, i.e., scraped data are curated to retain the best features (Das, Naik, Behera, Jaiswal, Mahato, and Rout 2020; Das, Naik, and Behera 2020b; Das, Naik, and Behera 2018).

Researchers have proposed numerous feature extraction algorithms for selecting the best features to perform data analytics (Das, Dey, and Balas 2019; Dey, Das, Naik, and Behera 2019) and various machine learning tasks like classification (Das, Naik, and Behera 2020c), clustering (Das, Naik, and Behera 2020a; Das, Jena, Nayak, Naik, and Behera 2015), forecasting (Rout, Jena, Rout, and Das 2020), and decision making (Dey, Ashour, Kalia, Goswami, and Das 2018).

For the notion that motivated this work, we use Google Scholar, Semantic Scholar, Scopus, Web of Science, and Microsoft Academic data to provide customizable recommendations for individuals. The aim is to support the research community by recommending the most suitable article from a curated list of publications for their domain of interest. For this drive, we developed a method to assess the publication attainment from a set of researchers who are involved in the identical research domain. The method comprises a scraper to acquire the required data. Figure 5.1 provides an overview of the tool.

Our proposed method uses a neural network to embed curated documents into a vector space by encoding the textual content of each document. We then select the nearest neighbors of a seed document as candidates and re-rank them using a second model

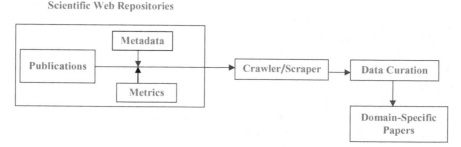

FIGURE 5.1   Outline of the data scraping.

trained to differentiate between cited and non-cited citations. Unlike existing works, our proposed model embeds newly published domain-centric documents in the same vector space used to identify journals similar to candidate journals based on their text content, obviating the need to retrain the models to include newly published journals.

The rest of this chapter presents mainstream challenges to scrape data from scientific web archives and plausible solutions to overcome these challenges. We also discuss pre-processing techniques used to handle this scraped data and build a recommender system to suggest suitable papers.

5.2 LITERATURE SURVEY

The scientific paper recommendation is a task that aims to find and recommend the most relevant papers from a large pool, given a domain of interests (Zhao, Wu, and Liu 2016). The facility to automatically filter a broad set of documents and find those documents that are most associated with one's research interest has its benefits. With the increasing amount of publications, many of them in web archives, it is challenging to keep track of the latest research, even if it is within one's area of interest or domain. With the timeliness of data becoming all the more critical, it is also necessary for a paper to reach researchers with minimal delay. In this work, we consider algorithms for curating integrated data based on rules and recommending a focused set of scientific articles from a particular domain.

The most common recommendation method employed in many applications like E-commerce, tourism, and entertainment websites are content and collaborative-based filtering. Existing works shows that these filtering techniques are also used for recommending research articles. Content-based filtering constructs the user report based on the user's reading pattern and recommends similar articles that best match the individual's reading pattern. In existing methods, user profiling is generally built by considering the importance of keywords; however, it is inadequate to model the individual's preference. To enhance the preference semantics, numerous methods have been proposed, such as the label-enriched approach (Guan, Wang, Bu, Chen, Yang, Cai, and He 2010) and the ontology-expansion approach (Zhang, Ni, Zhao, Liu, and Yang 2014). In the case of a collaborative filtering method, like-minded academic groups are identified first, and recommendations are generated based on their similar interests. But, the key concern in collaborative filtering is how to compute an individual's similarity. Davoodi et al. and Drachsler et al. have used data from various resources like e-mail logs, co-authors, references, and social media to analyze and to find the similarity among individuals (Davoodi, Afsharchi, and Kianmehr 2012; Drachsler, Hummel, and Koper 2008).

Many researchers have considered elements, such as domain knowledge, user background, learning targets, and cognitive patterns, apart from user preference when recommending resources (Zhao, Wu, and Liu 2016). Zhang et al. and Yang (2014) analyzed association rules between resources and courses, and recommendations were made for teaching resources. Tang et al. proposed a recommendation technique focusing on teaching features and combining the user's knowledge level and knowledge goals (Tang and McCalla 2004).

Some researchers have modeled domain knowledge as domain taxonomy, ontology, and concept networks. Liang et al. built a semantic network based on visited documents, where links between each semantic tree represent the inheritance relationship between concepts (Liang, Yang, Chen, and Ku 2008). Cantador et al. proposed a cluster-based paper recommendation method where each cluster in a semantic network represents users with similar preferences (Cantador and Castells 2006). De et al. developed an adaptive ontology based on the user's reading behaviors. The user's action was retrieved from the ontology, and recommendations were made based on the similar patterns observed (De Gemmis, Lops, Semeraro, and Musto 2015).

Meschenmoser et al. define web scraping as an automated technique to extract and retrieve targeted web data at range (Meschenmoser, Meuschke, Hotz, and Gipp 2016). A variety of tools and interfaces to build personalized scrapers, as well as customizable well equipped scraping frameworks, exist. Glez-Peña et al. and Haddaway et al. present ample summaries of frameworks and tools for various extraction tasks, namely DataToolBar, Helium Scraper, Screen Scraper, and FMiner (Glez-Peña, Lourenço, López Fernández, Reboiro-Jato, and Fdez-Riverola 2013; Haddaway 2015). But, there are very few scrapers for mining scientific records and bibliographic data. Smith-Unna et al. recommends a ContentMine framework that allows building personalized tools and other data mining elements (Smith-Unna and Murray-Rust 2014). Tang et al. propose an Aminer framework that collects and integrates heterogeneous social network data from many web data sources for researchers (Tang, Zhang, Yao, Li, Zhang, and Su 2008). But, the framework provides no provision for personalized content mining.

5.3 CONTENT-BASED RECOMMENDATION SYSTEM FOR DOMAIN-SPECIFIC PAPERS

In this chapter, to recommend a research article to the researchers working in a particular domain, we use a content-based recommendation system. The inspiration for journal paper recommendation is adopted from Content-Based Citation Recommendation proposed by Bhagavatula et al. (2018), which they use to recommend citations to an academic paper draft. The workflow of the proposed compendium network is presented in Figure 5.2, and each step is discussed in detail.

5.3.1 SCRAPING AND DATA INTEGRATION (CHALLENGES AND SOLUTIONS FOR DATA COLLECTION)

Commonly encountered obstacles when collecting data from scientific web archives through scraping can be categorized as follows (Meschenmoser, Meuschke, Hotz, and Gipp 2016):

1. Limitations on the size of the query results;
2. Dynamic contents;
3. Access limitations.

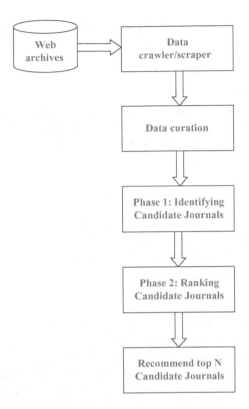

FIGURE 5.2 Workflow of DICN.

We will report on these obstacles and suggest approaches (solutions) to overcome the same in the following subsections.

5.3.1.1 Limitations on the Size of the Query Results

5.3.1.1.1 Fixed Limits

For any relevant query, scientific web archives often produce a fixed number of results. This fixed limit is suitable for people trying to retrieve a ranked item that is interactive, since a ranked retrieval system will usually revert to related outcomes within the first top-ten ranks. In contrast, the researcher will likely improve the search query if the desired information is not retrieved.

Before developing any scraper code, one has to inspect all the utilities reserved to retrieve any content from the web archives. This comprises spotting upper limits for data counts and examining all URL parameters. These imposed restrictions may vary for different item types; hence, it is recommended to investigate diverse data types.

For example, the search outcome for retrieving author details could be lower than the search results of scientific publications. Hence, the developer needs to design and develop the best setting for scraping a particular repository.

We have developed different scrapers for data collection and pipelining one scraper output as input to another scraper for efficient data collection. For example, one could scrape a URL of most cited and recent publications in a particular domain list and use the URL to download PDFs and then use the result sets to extract the authors' and co-authors' details.

5.3.1.1.2 Pagination

Apart from fixed limits on result sets, web archives usually use pagination for itemizing the contents in the query response set. In a standard pagination technique, the interface divides the resultant query response into many pages, each individual page displaying a preset amount of contents.

The presence of the pagination constraints may significantly affect a scraper's effectiveness. Few web archives use pagination constraints that allow web crawlers to ingress their data efficiently. For example, if the query outcomes identify the pages with consecutive numbers that are used for pagination, a web crawler can effortlessly retrieve the result set. If the item sub-lists are defined by other keywords like "start," "limit," and "end," web crawlers may even crawl and retrieve objects at random locations by posting a query.

5.3.1.2 Dynamic Contents

Web pages generally depend on the data that is dynamically and progressively updated using JavaScript. Dynamically loaded data throw an extreme challenge to scrapers in terms of complexity and run time performance.

Automated browsing through tools and frameworks by simulating a regular page visit are useful but are not very efficient for dynamic web content. For example, Selenium, a framework with WebDriver packages, supports and enables all modern browser automation for different programming languages. The drawback of this is that automated page visits decline a crawler's function and might not be entirely consistent in every set-up. Another approach for retrieving dynamic content is that the programmer needs to have prior knowledge (prerequisite) of how data is updated to the current page. There exist two methods to add data to a Hypertext Markup Language (HTML) page.

1. Hidden containers: Contents are exposed using JavaScript if some front end actions have been prompted, i.e., HTML code usually comprises the entire material once the page has loaded. Thus, this kind of dynamic data poses denial hindrance to scrapers.
2. Back end containers: Contents are dynamically retrieved by requesting materials from the server's back end if positive actions are prompted at the front end. In this setting, the data is not instantly accessible to a scraper. The JavaScript should be analyzed by a developer to explain which of the two methods should be used.

A researcher can inspect network logs and recognize URLs and parameters that are handled during POST and GET requests to access dynamic content. Based on the server's composition, the Web crawler will call the recognized URLs and processes the query.

5.3.1.3 Access Limitations

5.3.1.3.1 Masked URL Parameters

Typically, web repositories employ non-sequential URL parameters to obstruct content mining, for example, page identifiers with randomly selected characters. Usually, web repositories use identifiers that consist of 12 characters with lower and upper case letters, digits, a dash, and an underscore, which would result in 39* 1020 possible pages. In the case of scientific web archives, a considerable amount of these identifiers may not redirect to specific pages. Hence, crawling pages in the above conditions in the case of web repositories require other ways to acquire the identifiers. An alternative method that the web archives follow to obstruct the subsequent pages in a paginated environment is by creating reliance among the URL of the succeeding page and to the content of the current page. Scientific web archives use an "after item" factor that typically represents the identifier of a data element in the current page and that has to be specific to access the succeeding pages. In this case, the efficiency of the scraper significantly gets decreased as each page must be parsed to analyze the subsequent page. Directly accessing the arbitrary sub-lists using one request becomes hard in these cases.

5.3.1.3.2 Robot Recognition and Reverse Turing Tests

Numerous web archives devise various techniques to identify and block robots. Much research effort has been devoted to incorporate such identification techniques, which usually depend on machine learning algorithms. Balla et al. propose a decision tree based on ID3 (Balla, Stassopoulou, and Dikaiakos 2011). Bomhardt et al. recommend neural networks (Bomhardt, Gaul, and Schmidt-Thieme 2005), and Lu and Yu employed Hidden Markov Models for this purpose (Lu and Yu 2006).

If the identification approaches suspect an automated entrée effort, they may activate opposing Turing tests, which insist on specific user interactions to refacilitate entrée to the web resource, for example, static text CAPTCHAs, audio or video CAPTCHAs, image classification, and Google's "reCAPTCHA." To prevent being categorized as a robot and to avoid denial of access, a scraper can devise the below-listed approaches.

5.3.1.3.3 Changing the Content of Request Headers

A scraper could change user agents by simulating a user agent's identity for different requests.

5.3.1.3.4 Selecting Appropriate Cookie Settings

Depending on the scraping condition, the choice of whether to support or restrict cookies essentially needs a thorough study of the aimed-policy, because enabling cookies can help a scraper to seem more human.

5.3.1.3.5 Requests and Different Time Intervals

The interval between a series of requested URLs can be varied by choosing random time intervals or by deriving statistical data of user interactions. However, a trade-off among delaying requests and a scraper's efficiency has to be matched by developers.

5.3.1.3.6 *Altering the IP Address*

By using services such as "The Onion Router," requests with a different IP address can be used to send a query. The modification of the Internet Protocol address can happen at fixed or at arbitrarily selected time intervals, or when the queried server outputs errors, which generally indicates that the scraper was denied access.

We will alter the user operator settings, randomize the entities to scrape, restrict cookies, and alter IPs. Different time intervals among queries is not considered since the scraper developed in this chapter is acceptable for our use, and deferring queries might maximize the processing time.

5.3.2 DATA CURATION

Scraped and integrated data sets from multiple sources often contain missing and duplicate data that must be pre-processed to avoid deceptive learning processes leading to an undesirable set of recommendations. Our pre-processing stage involves transforming crawled web URLs into a suitable form to be able to deliver high-quality data for recommending domain-specific scientific papers. More specifically, we performed several pre-processing tasks, including data integration, named-entity recognition, handling missing values, and duplicate values (Shobha and Nickolas 2020).

The web crawled data (single day data) from different scientific web archives like Google Scholar, Semantic Scholar, Scopus, Web of Science, and Microsoft Academic includes the URLs of 4,681 papers. These URLs are the link for conference papers, journal papers, posters, books, and review articles. These URLs are further pre-processed to exclude books, reviews, conferences, and posters. Only 2,521 URLs were re-scraped to retrieve the entire document. Other pre-processing steps that were carried out are explained below.

1. **Handling irrelevant fields:** Several fields in the scraped raw data sets are not needed for recommendation purposes, such as GSrank, CitesPerYear, and Age of the paper. These fields are removed.
2. **Handling duplicate fields:** During the scraping process, fields with different names but having the same values have been collected. Since each web repository maintains its style of metadata about scientific papers, the chances of having the same values with different field names are high. When the integration of scraped data from different scientific web repositories is done, the presence of duplicate fields may contribute to high dimensional data. For example, fields like Estimated Citation Count (ECC) and Cites hold the same values forming duplicates. Hence, these duplicate fields are removed.
3. **Records with missing values:** There are several instances in the scraped data of missing values in different fields of articles, such as URL, DOI, and ISSN. These missing instances can be handled by human intervention by searching for it manually.
4. **Handling duplicate records:** As the developed crawler and scraper are domain-specific, they scrape data based on the domain name, for example, data imputation, handling missing values, or missing data handling. Scraped

data may or may not have a unique identifier like a DOI. Without a unique identifier, it may be hard to determine whether two records are similar or not in one phase. Hence, we consider various fields simultaneously or phase by phase to handle duplicate records. The following deliberations are made in this work to remove duplicate records based on the fields available after integrating data from multiple web archives:

a. Identifying duplicate records based on DOI.
b. A YEAR field (Publication/Accepted) can be utilized to mark the duplicate elements.
c. A fusion of fields like Title of paper + paging-info, ISSN+ Title of the article.
d. If de-duplication did not happen on the above rules, fields with lower accuracy, e.g., publisher name or author name, are used.

5.3.2.1 The Complexity of the Integration Operation

The simple and basic approach to handling duplicates in "R" records after integrating scraped data from multiple web archives is to compare record R_i with R_{i+1} where (i = 1, 2, 3, ..., n, where n = total number of records). This approach involves R^2 comparisons. Our work aims at integrating results of five web repositories that have three different queries (e.g., queries related to domain-specific publications, i.e., data imputation, handling missing values, missing data handling) resulting in 5 * 3 = 15 different data sets. To handle duplicates in these scraped data, a multiple join procedure has to be carried out. For every query, if the web repositories return "R" records, the total number of analogies to remove the duplicates relies upon the rate of duplicates among the list of records of the web archive outcomes in the lists.

Overall, the worst-case time complexity to carry out de-duplication is "$D * R^2$", where "D" is constant when there are no duplicate records.

Considering the time complexity of this naive method, the proposed de-duplication method uses an alternative method by reducing the number of comparisons by creating blocks based on YEAR of publication, so that only records of the specific block need to be compared, resulting in a lower number of comparisons.

The proposed method recommends a new document to the user by finding nearby candidates after embedding a seed document into the linear space. As a second step, re-ranking of the documents is done using a model that is competent enough to distinguish between cited and uncited references. The entire process of encoding and embedding the textual content of each document into a linear space is done through a neural network. We articulate paper recommendations as a ranking problem. Given a seed document S_d and a large corpus of integrated papers, the task is to rank documents so that the document with a higher rank could be recommended to the user working in a particular domain. The advantage of crawling and scraping the documents that are relevant to a specific domain is that the number of published papers in the scientific web archives can be significantly large and that it is computationally expensive to rank each document as a candidate reference concerning seed document S_d. Hence, we recommend citations on scraped and integrated domain-specific data.

5.3.3 PHASE 1: IDENTIFYING CANDIDATE JOURNALS

In this phase, our goal is to recognize a set of candidate journals that are similar to seed documents, S_d. Using a trained neural network, we first embed all the curated documents into a linear space such that domain-centric documents tend to be closer. Since the projection of a document is independent of the seed document, the entire curated documents need to be projected only once and can be reused for subsequent queries. Then, we project each seed document S_d to the same linear space and identify its "k" nearest neighbors as candidate references. For the nearby candidates, both outgoing and incoming citations are considered (Strohman, Croft, and Jensen 2007). The outcome of this phase is a list of documents S_i and their corresponding scores $sim(S_d, S_i)$, described as a cosine similarity between S_d and S_i in the embedding space. This phase yields a manageable number of candidate journals, making it practical to score each candidate Di by feeding the pair (S_d, D_i) into another neural network in Phase 2, trained to discriminate whether the journal should be recommended or not. An overview of this phase is represented in Figure 5.3.

A supervised learning model is used to project the contents of document D to a dense embedding. Each textual field of D is represented as a bag-of-words, and feature vectors are computed as given in Equation 5.1 (Bhagavatula, Feldman, Power, and Ammar 2018):

$$FV_{D[Title]} = \sum_{t \in D[Title]} d_t^{mag} \frac{d_t^{dir}}{\left\| d_t^{dir} \right\|^2} \tag{5.1}$$

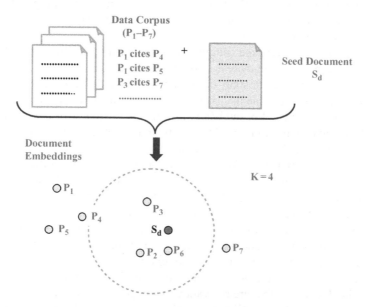

FIGURE 5.3 An overview of Phase1: All the papers in the corpus (P_1–P_7) in this example are projected into a linear space in addition to the seed document S_d. Nearest neighbors of S_d are chosen as the nearest *candidates*. Considering k = 4, P_2, P_6, P_3, and P_4 are selected as the nearest candidates. P_7 is also considered as the nearest candidate as it is cited in P_3.

where d_t^{dir} is a dense direction embedding and d_t^{mag} is a scalar magnitude of word type t. The weighted average of fields is computed after normalizing the representation of each field to get the document embedding, D_e. The corresponding equation for normalizing is given in Equation 5.2. Here, we use the title, abstract, and keywords field of a document D:

$$D_e = \lambda^{Title} \frac{FV_{D\{Title\}}}{\left\| FV_{D\{Title\}} \right\|^2} + \lambda^{Abstract} \frac{FV_{D\{Abstract\}}}{\left\| FV_{D\{Abstract\}} \right\|^2} + \lambda^{Keywords} \frac{FV_{D\{Keywords\}}}{\left\| FV_{D\{Keywords\}} \right\|^2} \quad (5.2)$$

where λ^{Title}, $\lambda^{Abstract}$, and $\lambda^{Keywords}$ are scalar model parameters.

The parameters of embedding mode, that is, $\lambda*$, d_*^{mag}, and d_*^{dir}, are learned using a training set T of triplets $< S_d, D^{cited}, D^{notcited} >$ where S_d is a seed document, D^{cited} is a document cited in S_d, and $D^{notcited}$ is a document not cited in S_d. Wang et al. describe in their work that the goal of the model training is to predict a high cosine similarity for the pair (S_d, D^{cited}) and a low cosine similarity for the pair (S_d, $D^{notcited}$) using the per-instance triplet loss (Wang et al. 2014). The equation representing the loss function is given in Equation 5.3.

$$\text{Loss} = \max \left(\alpha + s\left(S_d, D^{notcited} \right) - s\left(S_d, D^{notcited} \right), 0 \right) \quad (5.3)$$

where $s(S_i, D_j)$ is defined as the cosine similarity between document embeddings cos-sim (D_{ei}, D_{ej}). Here α is considered as the hyper-parameter of the model and tuned to get the best results. Choosing a positive sample for training is straightforward, any pair of (S_d, D^{cited}) can be selected, where a document S_d in the training set cites D^{cited}. But, a careful selection of the not-cited sample is needed for training the model to show better performance. Here, the documents that are not cited in a seed document but that are near to it in the embedding space *are* chosen as negative samples to train the model.

5.3.4 PHASE 2: RANKING CANDIDATE JOURNALS

This phase aims at training the model, which takes a pair of documents (S_d, D_i) and estimates the probability as to whether D_i should be recommended or not.

A vital goal of this work is to evaluate the viability of recommending journals without using metadata. $FV_{D[field]}$, a dense feature vector, is calculated for different fields (title, abstract, and keywords) of each document.

For the document under consideration, a subset of word types in the title, abstract, and keywords are identified by conducting intersection, and the sum of their scalar weights are computed as an additional feature, for example, $\Sigma_{te \cap abstract} d_t^{\cap}$. Along with this, we also consider the log of the outward connection (the number of other documents cited) of D_i, that is, $\log(D_{i[out\text{-}citations]})$. Lastly, the cosine similarity between S_d and D_i in the embedding space, that is, $cos\text{-}sim(D_{ei}, D_{ej})$, is used. An overview of this phase is represented in Figure 5.4.

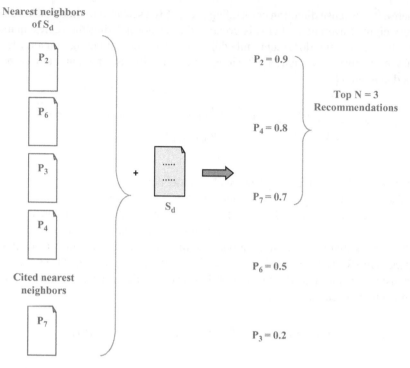

FIGURE 5.4 In this phase, each pair (S_d, P_2), (S_d, P_6) (S_d, P_3) (S_d, P_4), and (S_d, P_7) are scored separately to re-rank the papers; the top-three papers P_2, P_4, and P_7 are recommended.

Architecture: In our work, a supervised "feed forward" neural network is used to rank the documents. This network architecture has three layers, namely, two exponential linear unit layers, and one sigmoid layer. The model is as shown in Figure 5.5. This architecture is used to compute the cosine similarity between the embedding of S_d and S_i for textual and categorical fields. Then, the cosine similarity scores are concatenated with numeric features, the weighted sum of intersection words, and two dense layers with the exponential linear unit, a non-linear activation function. The final layer is an output layer with a non-linear sigmoid function, which estimates the probability that S_i is recommended or not. The output layer is defined as:

$$s(S_i, D_i) = FeedForward(O),$$

$$O = [G_{title}, G_{authors}, G_{abstract}, G_{keywords}; \cos - sim(D_{ei}, D_{ej});$$

$$\sum_{te \cap abstract} d_t^{\cap}; D_i[out - citations]G_{field} = \cos - sim\left(FV_{D_{ei[field]}}, FV_{D_{ej[field]}}\right) \quad (5.4)$$

Training the model: The parameters used in Phase 2 are d_*^{mag}, d_*^{dir}, $d_*^{\cap}$, and the parameters of the three dense layers in the supervised neural network. The triplet loss function defined in Equation 5.3 is used to train these parameters and redefine the

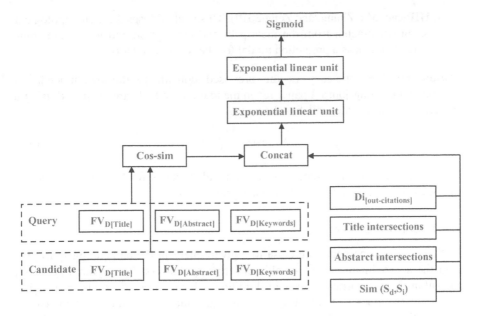

FIGURE 5.5   Architecture of a neural network.

similarity functions $s(S_i, D_i)$ as the neural network output set in Equation 5.4. During tests and while recommending the real-time query, we used this proposed model to recommend journal D_i with the highest $s(S_d, D_i)$ scores.

5.4 EXPERIMENTAL RESULTS AND DISCUSSIONS

In this section, we compare the experimental results of our proposed compendium network recommendation method with the following methods.

1. **Latent Dirichlet Allocation (LDA)** (Blei, Ng, and Jordan 2003): This method handles the content affinity among the topics that are obtained from the title of the papers. For the user's new seed document S_d, we use the text content of seed document S_d and the documents referenced by S_d to train the topic distribution for S_d.
2. **Collaborative Topic Regression (CTR)** (Wang and Blei 2011): This method combines matrix factorization and the probabilistic topic model, which studies user-item ratings and textual affinity between papers.
3. **User-Article based Graph Model with Tags (UAGMT)** (Cai, Cheng, Luo, and Zhou 2016): This method proposes new paper recommendation by considering the bipartite graph, which takes the readership, tag, content, and citation into account.
4. **MPRec** (Ma, Zhang, and Zeng 2019): This method trains the model using meta-path-based topological features and resolves the recommendation problem in heterogeneous information networks.

5. **HIPRec** (Ma, Zhang, and Zeng 2019): This method fuses different topological features extracted from the meta-path and meta-graph and uses these combined features as a prediction model for the recommendation.

Evaluation: The efficiency of the proposed domain-specific recommendation method and existing journal paper recommendation methods are evaluated using a prediction accuracy rate, which is defined as:

$$\text{Accuracy} = \frac{R_{sd}}{N_t} \tag{5.5}$$

where R_{sd} represents the number of recommended papers for the query researcher and N_t indicates the total number of tested journal papers.

5.4.1 CONFIGURATIONS

To find similar papers that are close to a given query in the embedding space, we used the approximate nearest neighbors (https://github.com/spotify/annoy) search algorithm. This algorithm works by building a binary tree of points in a higher dimension by choosing random hyper-planes that allow searching of nearby candidates in O(log n) time. To optimize various hyper-parameters, such as the size of hidden layers, learning rate, and regularization strength, we used the python library "hyperopt."

We evaluated the accuracy of the proposed DICN technique and the other existing techniques from the literature. From Figure 5.6, it can be seen that LDA and CTR performances are low compared to other methods. The reason behind this low performance is that LDA and CTR rely only on the title data of the paper, which is not sufficient to generate accurate recommendations. UAGMT maps the recommendation problem into a bipartite graph and solves the problem by considering text contents and citations. We may notice from Figure 5.6 that the performance of UAGMT is somewhat superior to that of LDA and CTR. From Figure 5.6 we can see by analysis that the proposed compendium network attains significantly improved accuracy

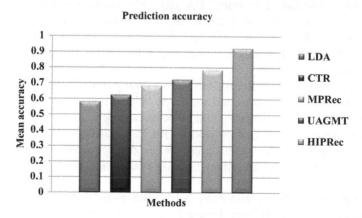

FIGURE 5.6   Prediction accuracy comparison of the proposed and existing methods.

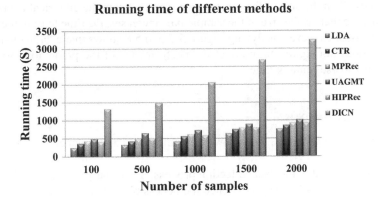

FIGURE 5.7   Prediction accuracy for different samples.

over all other models. This is because our proposed technique is constructed to handle the paper recommendation problem in the domain-specific environment.

Secondly, we assess the competence of the proposed DICN method with a varying number of samples. The outcomes are as shown in Figure 5.7. It can be seen from the figure that the prediction accuracy of the proposed method improves with the increase in the number of training samples. This is because, with the rise in the training sample size, better features can be learned, thus the learning model will be more significant. However, when the number of sample pairs reaches around 1000, the set size does not significantly affect the efficiency of the model. To conclude, an almost small share of training samples are enough for the proposed compendium network to learn the patterns effectively.

Figure 5.8 shows the running time of the proposed method and other existing methods with a varying number of samples given as input. It can be seen that there is a substantial rise in running time for the proposed method compared to existing

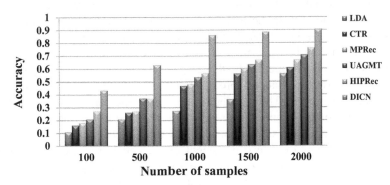

FIGURE 5.8   Running time comparison of the proposed and existing methods.

techniques when the size of the sample pairs grows. From the empirical results, one can conclude that, as the size of the sample pair increases, the time to train the model also increases. However, analyzing Figure 5.7 and 5.8 shows that a large quantity of samples does not significantly affect the performance of the prediction model but results in a worse running time.

5.4.2 Result Analysis

From the empirical results, the following conclusions can be made.

1. As explained in the data collection process, it is usually hard to get all the text data of even a few papers. Hence the accessible data of the papers are somewhat partial; for example, when only the title of the paper can be used, then the content-based recommendation achieves poor outcomes. In contrast, the computational complexity will be very high if all the text data of the paper are used for solving the recommendation problem.
2. Most of the existing graph-based methods do not consider the semantic meaning underlying the networks; they handle the recommendation problem using homogeneous or bipartite networks.
3. Compendium network-based recommendation is influential and it encapsulates much more vibrant and more complex semantic meaning than simple graph-based methods. Recommendation approaches that incorporate domain-specific data perform better with less computational cost compared to other methods.

5.5 CONCLUSION AND FUTURE WORK

In this chapter we have proposed a neural network and the nearest neighbor based method to recommend domain-centric research articles for researchers. To do so, we built a knowledge network by scraping, integrating, and pre-processing domain-centric data from multiple web repositories, like Google Scholar, Semantic Scholar, Scopus, Web of Science, and Microsoft Academic data. Mainly, we have suggested a method to recommend domain-centric research articles for researchers based on their domain of interest. In the proposed compendium network, the user's domain knowledge and his or her research interests are used to find out the knowledge gap, and the compendium network aims to recommend domain-centric papers that can help bridge that gap. The method firstly scrapes the domain-centric data based on the domain corpus. Then the data is thoroughly pre-processed to make it suitable for the recommendation process. The contributions of this chapter are as follows. In the first phase, curated domain-centric documents are embedded into a linear space using a neural network such that a similar domain-centric document tends to be closer. Then the nearest neighbors of a seed document are selected as candidates, which are re-ranked using a neural network trained to assign a probability rank to differentiate whether the document should be recommended or not. The empirical outcomes show that the proposed domain-centric-based method gives the best overall performance in comparison to the state-of-the-art methods. The domain-centric data reduce the search space and help in faster identification of the documents similar to the seed document and thus improve the recommendation by reducing the computational cost. Regarding future work, the proposed compendium network method could be automated and deployed in a distributed environment to recommend domain-centric research articles.

REFERENCES

Balla, A., A. Stassopoulou, and M. D. Dikaiakos. 2011. Real-time web crawler detection. In *2011 18th International Conference on Telecommunications*. IEEE, Ayia Napa, Cyprus, pp. 428–432.

Basu, C., Hirsh, H., Cohen, W. W., and Nevill-Manning, C. 2001. Technical paper recommendation. A study in combining multiple information sources. *Journal of Artificial Intelligence Research*, 14, 231–252.

Bhagavatula, C., S. Feldman, R. Power, and W. Ammar. 2018. Content-based citation recommendation. arXiv preprint arXiv:1802.08301.

Blei, D. M., A. Y. Ng and M. I. Jordan. 2003. Latent Dirichlet allocation. *Journal of Machine Learning Research* 3(Jan):993–1022.

Bomhardt, C., W. Gaul, and L. Schmidt-Thieme. 2005. Web robot detection pre-processing web log files for robot detection. In *New Developments in Classification and Data Analysis*. Springer, Berlin, pp. 113–124.

Cai, T., H. Cheng, J. Luo, and S. Zhou. 2016. An efficient and simple graph model for scientific article cold-start recommendation. In *International Conference on Conceptual Modeling*. Springer, Berlin, pp. 248–259.

Cantador, I. and P. Castells. 2006. Multilayered semantic social network modelling by ontology-based user profiles clustering: Application to collaborative filtering. In *International Conference on Knowledge Engineering and Knowledge Management*. Springer, pp. 334–349.

Das, H., N. Dey, and V. E. Balas. 2019. *Real-Time Data Analytics for Large Scale Sensor Data*. Academic Press, London.

Das, H., A. K. Jena, J. Nayak, B. Naik, and H. Behera. 2015. A novel pso based back propagation learning-MLP (PSO-BP-MLP) for classification. In *Computational Intelligence in Data Mining*. Springer, Singapore, Vol. 2, pp. 461–471.

Das, H., B. Naik, and H. Behera. 2018. Classification of diabetes mellitus disease (DMD): A data mining (DM) approach. In *Progress in Computing, Analytics and Networking*. Springer, Singapore, pp. 539–549.

Das, H., B. Naik, and H. Behera. 2020a. An experimental analysis of machine learning classification algorithms on biomedical data. In *Proceedings of the 2nd International Conference on Communication, Devices and Computing*. Springer, Singapore, pp. 525–539.

Das, H., B. Naik, and H. Behera. 2020b. A hybrid neuro-fuzzy and feature reduction model for classification. *Advances in Fuzzy Systems* 2020:1–15.

Das, H., B. Naik, and H. Behera. 2020c. Medical disease analysis using neuro-fuzzy with feature extraction model for classification. *Informatics in Medicine Unlocked* 18.

Das, H., B. Naik, H. Behera, S. Jaiswal, P. Mahato, and M. Rout. 2020. Biomedical data analysis using neuro-fuzzy model with post-feature reduction. *Journal of King Saud University-Computer and Information Sciences*.

Davoodi, E., M. Afsharchi, and K. Kianmehr. 2012. A social network-based approach to expert recommendation system. In *International Conference on Hybrid Artificial Intelligence Systems*. Springer, Berlin, pp. 91–102.

De Gemmis, M., P. Lops, G. Semeraro, and C. Musto. 2015. An investigation on the serendipity problem in recommender systems. *Information Processing & Management* 51(5):695–717.

Dey, N., A. S. Ashour, H. Kalia, R. Goswami, and H. Das. 2018. *Histopathological Image Analysis in Medical Decision Making*. IGI Global, Hershey, PA.

Dey, N., H. Das, B. Naik, and H. S. Behera. 2019. *Big Data Analytics for Intelligent Healthcare Management*. Academic Press, Boston, MA.

Drachsler, H., H. Hummel, and R. Koper. 2008. Identifying the goal, user model and conditions of recommender systems for formal and informal learning. *Journal of Digital Information* 10(2):4–24

Glez-Peña, D., A. Lourenço, H. López-Fernández, M. Reboiro-Jato, and F. Fdez-Riverola. 2013. Web scraping technologies in an API world. *Briefings in Bioinformatics* 15(5):788–797.

Guan, Z., C. Wang, J. Bu, C. Chen, K. Yang, D. Cai, and X. He. 2010. Document recommendation in social tagging services. In *Proceedings of the 19th International Conference on World Wide Web*, Raleigh, pp. 391–400.

Haddaway, N. R. 2015. The use of web-scraping software in searching for grey literature. *Grey Journal* 11(3):186–190.

Liang, T.-P., Y.-F. Yang, D.-N. Chen, and Y.-C. Ku. 2008. A semantic-expansion approach to personalized knowledge recommendation. *Decision Support Systems* 45(3):401–412.

Lu, W.-Z. and S.-Z. Yu. 2006. Web robot detection based on hidden markov model. In *2006 International Conference on Communications, Circuits and Systems*. IEEE, Guilin, Vol. 3, pp. 1806–1810

Ma, X., Y. Zhang, and J. Zeng. 2019. Newly published scientific papers recommendation in heterogeneous information networks. *Mobile Networks and Applications* 24(1):69–79.

Meschenmoser, P., N. Meuschke, M. Hotz, and B. Gipp. 2016. Scraping scientific web repositories: Challenges and solutions for automated content extraction. *D-Lib Magazine* 22(9/10).

Pan, C. and W. Li. 2010. Research paper recommendation with topic analysis. In *2010 International Conference on Computer Design and Applications*. IEEE, Qinhuangdao, China, Vol. 4, pp. 4–264.

Rout, M., A. K. Jena, J. K. Rout, and H. Das. 2020. Teaching–learning optimization based cascaded low-complexity neural network model for exchange rates forecasting. In *Smart Intelligent Computing and Applications*. Springer, Berlin, pp. 635–645.

Shobha, K. and S. Nickolas. 2020. Integration and rule-based pre-processing of scientific publication records from multiple data sources. In *Smart Intelligent Computing and Applications*. Springer, Berlin, pp. 647–655.

Smith-Unna, R. and P. Murray-Rust. 2014. The content mine scraping stack: Literature-scale content mining with community-maintained collections of declarative scrapers. *D-Lib Magazine* 20(11/12).

Strohman, T., W. B. Croft, and D. Jensen. 2007. Recommending citations for academic papers. *SIGIR* 7:705–706.

Tang, J., J. Zhang, L. Yao, J. Li, L. Zhang, and Z. Su. 2008. Arnetminer: Extraction and mining of academic social networks. In *Proceedings of the 14th ACM SIGKDD International Conference on Knowledge Discovery and Data Mining*. ACM, pp. 990–998.

Tang, T. and G. McCalla. 2004. Beyond learners interest: Personalized paper recommendation based on their pedagogical features for an e-learning system. In *Pacific Rim International Conference on Artificial Intelligence*. Springer, Berlin, pp. 301–310.

Wang, C. and D. M. Blei. (2011). Collaborative topic modelling for recommending scientific articles. In *Proceedings of the 17th ACM SIGKDD International Conference on Knowledge Discovery and Data Mining*. ACM, San Diego, CA, pp. 448–456.

Wang, J., Y. Song, T. Leung, C. Rosenberg, J. Wang, J. Philbin, B. Chen, and Y. Wu. 2014. Learning fine-grained image similarity with deep ranking. In *Proceedings of the IEEE Conference on Computer Vision and Pattern Recognition*. San Juan, Puerto Rico, pp. 1386–1393.

Zhang, H., W. Ni, M. Zhao, Y. Liu, and Y. Yang. 2014. A hybrid recommendation approach for network teaching resources based on knowledge-tree. In *Proceedings of the 33rd Chinese Control Conference*. IEEE, Nanjing, China, pp. 3450–3455

Zhao, W., R. Wu, and H. Liu. 2016. Paper recommendation based on the knowledge gap between a researcher's background knowledge and research target. *Information Processing & Management* 52(5):976–988.

6 Forecasting Air Quality in India through an Ensemble Clustering Technique

J. Anuradha, S. Vandhana, and Sasya I. Reddi

CONTENTS

6.1 INTRODUCTION

Air pollution has been a major cause of various diseases in recent years. The ambient quality of air (outdoor air quality) manifested worldwide deaths of 4.2 million in the year 2016. Prolonged exposure to particulate matter (PM) of diameter less than 2.5 microns can prove to be fatal (WHO 2018). It causes cardiovascular and respiratory diseases and cancers. It is one of the incremental problems affecting all countries in the world.

Rather than affecting human health directly, air pollutants can cause long-term damage to the environment which results in climate change and an indirect threat to

113

health and well-being. Besides exposure to PM, health can be drastically affected by exposure to ozone (O_3) as well. One of the major factors of asthma morbidity and mortality is ozone, which can cause reduced lung functioning, lung cancer, and other breathing problems. The air pollutants NO_2 and $PM_{2.5}$ have several correlated activities as shown by Dai, Luo, Luo, Qin, and Peng (2014). One of the main components of nitrate aerosols is NO_2. Epidemiological studies have shown that NO_2 has also been linked to reduced lung function. In a few cities in North America and Europe, the concentration levels of NO_2 are being monitored regularly. SO_2 has been known to aggravate chronic bronchitis and asthma, make people more vulnerable to respiratory tract infections, and cause respiratory tract inflammation and eye irritation. On the days when the concentration levels of SO_2 are higher, the number of cases of cardiac diseases or mortality is higher, as stated by the WHO (2018).

The National Wildlife Federation of the USA in the late 1960s developed the Environmental Quality Index (EQI), of which the Air Quality Index plays an integral part. Many researchers have concentrated on ambient air quality and applied the methods of classification, clustering, prediction, and forecasting for the well-being of humans. Upon clustering or grouping, one can find the exact shape of data or the patterns that it is falling in and so one can take the control measures that reduce the affect on human health in the future.

Various data mining research activities involve clustering as a key step, as it aims at separating information into clusters or classes on the basis of a certain measure of similarity. The final objective is to place the data points in the same cluster which are similar to each other and points that differ in separate clusters. It is well known that different clustering methods when applied to the same set of data can discover different patterns. This is because, due to the optimization of different criteria, each algorithm has its own bias. Another challenge of clustering is related to the validation of the results when ground truth is not available.

Clustering ensembles have appeared in recent years as a strategy for solving some of the clustering problems, as shown by Ghosh and Acharya (2011) and Strehl and Ghosh (2012). An ensemble clustering method is described by its two major components:

- Generation of diverse partitions;
- Combining partitions into a final clustering using a consensus function.

The same algorithm with different initializations or the same data with different bootstrap samples or algorithms make a clustering ensemble system. A clustering ensemble offers a solution to various challenges that are inherent to clustering. By using consensus through several clustering tests, more reliable and robust solutions can be provided. Different data samples may induce variance which can lead to spurious structures, which also emerge due to various biases and can be averaged out by the consensus function.

A subspace clustering can also be viewed as a weighted cluster array, in which each cluster represents the importance of its features using a weight vector. In the subspace clustering ensemble, the clusters and the weight vectors given by the base subspace clusters are used in the consensus function. The final consensus clustering can be improved by evaluating the relevance of the base clustering (and the assignments of weights accordingly) as shown by Li, Ding, and Jordan (2007). To the best of our knowledge, Nock and Nielsen (2006) were the first to explore how to use object related weights in the ensemble clustering method.

Researchers analyzed the advantages of calculating objects combined with advanced techniques in iterative clustering approaches, such as a probabilistic model, a K-harmonic mean, and a K-harmonic mean for clustering with a boosting technique, as shown by Hamerly and Elkan (2012), Topchy, Jain, and Punch (2014), and Zhang, Hsu, and Dayal (2000). Empirical observations suggest that objects that are difficult to cluster should be given more weights.

Centroid-based clustering methods involve moving the centers of the clusters to object regions where it is difficult to identify the membership of these objects in a cluster. In boosting, the areas around the difficult points are densified because the distribution of data is biased by the weights. Thus, the centers of the clusters are moved towards the weight-modified distribution modes. Using Bregman divergence, clustering was formulated as a constrained minimization by Nock and Nielsen (2016). Objects that are hard to cluster are recommended to be given large weights. A few algorithms based on weighted versions were introduced and the advantages of using boosting clustering techniques were analyzed.

The rest of the chapter is organized firstly in terms of related works on air quality prediction, ensemble modeling, and ensemble clustering. Then follows dataset description, methodology explanation of the ensemble consensus function and the METIS function, and finally the experimental results to supplement the proposed work.

6.2 RELATED WORKS

6.2.1 AIR QUALITY PREDICTION

Based on the geographical area, the urban emissions causing air pollution have been analyzed and mapped by Asgari, Farnaghi, and Ghaemi (2017). Apache Spark was used to study Tehran data from 2009 to 2013. In addition, the accuracy was predicted by Naive Bayes and logistic regression algorithms. In conclusion, the data was more accurately forecast by Naive Bayes in comparison to other machine learning algorithms that are used to identify unknown air quality groups. So strong results were obtained in this paper for Apache Spark.

In Zhu, Cai, Yang, and Zhou (2018), optimization and regularization techniques were used to predict the air pollutant values for the next day. The values of the different air pollutants that were predicted in the paper were that of sulfur dioxide, particle matter ($PM_{2.5}$), and ozone. Data from the two stations were used for prediction. The values for O_3 and SO_2 were predicted by one station and O_3 and $PM_{2.5}$ by the other. The data was modeled based on similarity and, for grouping, linear regression was used. The evaluation criteria was the root-mean-squared error (RMSE); however, the linear regression model failed to forecast or handle unforeseen events.

Gore and Deshpande (2017) studied the effects of air quality on health and the classification of the air quality index. For classification, they implemented a decision tree, J48, and Naive Bayes, and the results prove that the decision tree algorithm performs better than the rest with an accuracy of 91.9978%. However, this research has certain shortcomings as the dataset is limited and the decision tree cannot perform well on continuous variables and has issues with overfitting. A classification of air quality index was proposed using K-means algorithm. Again the dataset was limited, as the K-means technique was unfit for predicting the future values, as shown by Kingsy, Manimegalai, Geetha, Rajathi, Usha, and Raabiathul (2016).

The limitation of computational models for air quality is discussed by NASA Goddard Space Flight (2018). They proposed different machine learning techniques to forecast the concentration levels of O_3 in various countries. The dimensionality of the data is reduced in the pre-processing technique (sparse sampling and randomized matrix decompositions). A random forest regression technique is used for forecasting the next ten days. The researchers used only one pollutant (O_3) for future prediction and the data subsample size is small. Air pollution prediction using the dynamic neural network (DNN) approach was carried out on data generated by low cost sensors by Esposito, De Vito, Salvato, Bright, Jones, and Popoola (2016).

Deep learning techniques to predict the concentration levels of ozone in smart cities was proposed by Ghoneim et al. (2017). Deep learning using a feed forward neural network was used on the Aarhus city dataset. The model was compared with neural networks (NN) and support vector machine (SVM). This proved that deep neural networks accurately measured the pollution value.

Ozone concentration was studied in Tunisia by Ishak, Daoud, and Trabelsi (2017). Ozone concentration recorded at three stations was considered for prediction. A random forest and support vector regression were applied for prediction. Random forest was found to be more accurate than SVM.

6.2.2 ENSEMBLE MODELING

Ensemble modeling is a machine learning technique in which several base diverse learning models are used to predict the output. The objective of this approach is to decrease the variance and bias of the model, which reduces the generalization error and also provides stability to the model. In this way performance of the ensemble models are improved. As long as the base models are independent and diverse, the performance is always better. This technique uses the wisdom of collective results from base learners by applying any one of the aggregation methods mentioned in Figure 6.1. With this approach one can construct a strong learner out of weak learners by using one of the following approaches:

1. Bagging (reduce variance);
2. Boosting (reduce bias);
3. Stacking (improves prediction).

Figure 6.2 shows the construction of an ensemble model from various base learners. Here, the training samples are divided into multiple subsets of samples by random selection with replacement. This selection method is call *bootstrap aggregation* or *bagging* and involves repeated random resampling of training data (Breiman 1996). The sub-sample data may have complete features or a subset of features and each of

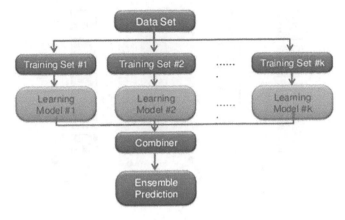

FIGURE 6.1   List of aggregation functions applied on ensemble model.

Rule	Fusion function $f(\cdot)$
Sum	$y_i = \frac{1}{L}\sum_{j=1}^{L} d_{ji}$
Weighted sum	$y_i = \sum_j w_j d_{ji}, w_j \geq 0, \sum_j w_j = 1$
Median	$y_i = \text{median}_j d_{ji}$
Minimum	$y_i = \min_j d_{ji}$
Maximum	$y_i = \max_j d_{ji}$
Product	$y_i = \prod_j d_{ji}$

FIGURE 6.2   Construction of ensemble model.

these samples are trained with a base learner. Finally the results from all the learners are aggregated. Decision trees and neural networks are unstable learners because even if the training data is changed slightly, the output may change. This variance in the results and the error it may cause is decreased by bagging.

The *boosting* method in ensembling uses a re-weighting approach for the samples during the training phase. This approach is capable of boosting the performance of the weak learners, several of which are involved iteratively during the training of a sample. Results from the different hypotheses are combined to form a strong learner. This was first introduced and later revised into the AdaBoost algorithm by Schapire (1999).

Initially, all samples are assigned uniform weights. Weak learners will train the weighted samples. From the results of the model, based on misclassification, the weights are reassigned to the sample. The weightage for the misclassified samples are increased and for the correctly classified samples it is reduced. Iteratively, in this manner, weak learners are trained. The results from all the base learners are aggregated. This is depicted in Figure 6.3. The boxes in the figure are proportionate to the weights of the sample. The tick and cross symbols represent the correct and incorrect classifications. Every base learner produces a hypothesis. Thus h_1, h_2, h_3, and h_4 are generated by the four weak learners trained on weighted samples. Finally, the hypotheses are aggregated to generate a single hypothesis. Samples are tested on this hypothesis.

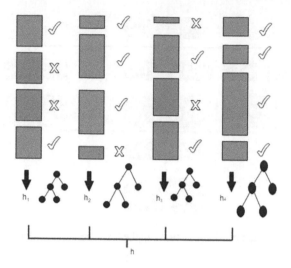

FIGURE 6.3   Re-weightage in the boosting model.

Stacking combines several base learners using a meta-classifier or regressor. Base learners are trained on a subset of samples. A meta-classifier/regressor works on top of the base learner. The procedure for stacking is given below.

The training set is split into two sets that are disjoint:

1. In one part, several base learners are trained;
2. In the other part, the base learners are tested;
3. Higher level learners are then trained using input as the predictions from the previous step and the output as the correct responses.

6.2.2.1 Variants of Ensemble Models

Variant approaches for ensemble methods are proposed using fuzzy, neural network, and some statistical approaches. Fuzzy instance weightage assignment for ensemble classification on data streams was proposed to identify the concept drift. This is an adaptive approach that uses a dynamic voting method. This method proposed by Dong et al. has the advantage of less computational cost with better accuracy. It is adaptable and can recognize concept drift.

A dynamic weighted neural network of ensemble models uses a bagging scheme. The dynamic weightage method uses integration of a neural network ensemble classifier. This method overcomes the performances of traditional integration methods (Li et al. 2007). The composite prediction output is obtained by combining various Long Short Term Memory (LSTM) models by dynamically adjusting the combining weights. By using the forgetting weight factor and past prediction errors the combining weights can be updated in a recursive and adaptive way (Choi and Lee 2018).

6.2.3 ENSEMBLE CLUSTERING

Now, moving on to the clustering techniques, many domains have proven that in comparison to individual classification techniques, a classifier ensemble is more accurate in most cases. This has initiated research work in the area of ensemble methods for clustering. Fred and Jain (2002) generated diverse base clustering results with different initialization of centroids in a K-means clustering algorithm. Similarity between samples is measured by the co-association matrix, which the K-means results are mapped onto. The work was extended by Kuncheva and Hadjitodorov (2014), in which a random number of clusters was chosen for each member that was a part of the ensemble clustering. A procedure for meta-clustering, consisting of two steps, was introduced by Zeng, Tang, Garcia-Frias, and Gao (2012). Initially, all the clusters are converted into a distance matrix. Next, the various distance matrices from the clustering are combined using a hierarchical clustering method, thus introducing a consensus clustering for computing. Hu (2008) generated combined clustering results by using a graph-based partitioning approach. Ayad and Kamel (2013) introduced a graph approach with vertices and edges. Data points are represented by vertices and when a certain number of nearest neighbors are shared between two data points, there exists an edge between the vertices. A random projection of data points is combined with a cluster ensemble by Fred and Jain (2002). Clustering is carried out using the expectation and maximization approach and an agglomerative hierarchical clustering is applied to obtain the concluding results. Greene, Tsymbal, Bolshakova, and Cunningham (2014) focused on generating different integration techniques for input clustering. The dataset used was medical diagnostic data. Base clustering was generated using fast, weak K-medoids and K-means clustering. The aggregation of results is given by a co-occurrence matrix, upon which hierarchical clustering schemes are applied for consensus ensemble clustering. Different hierarchical methods, such as single linkage, complete linkage, mean, median, and ward, were used to cluster the Dengue data (Vandhana and Anuradha 2018). The base clustering results can also be produced from various hierarchical methods.

The normalized mutual information between the clusters is maximized by Strehl and Ghosh (2012), combining the clustering results using a novel consensus function. The three heuristics represent the ensemble clustering as a hyper-graph. Each of the partitionings is represented as a hyper-edge. The three heuristics are: a meta-clustering algorithm (MCLA), a hyper-graph partitioning algorithm (HGPA), and a cluster-based similarity partitioning algorithm (CSPA). In CSPA, the inputs for clustering are converted into a binary similarity matrix. For each pair of points, the value is 1 if it belongs to the same cluster, 0 otherwise. A similarity matrix S is generated using the average of all the matrices. The results are re-clustered from S, with a graph-based partitioning approach. The generated similarity graph consists of vertices corresponding to data, and edges represent the weight of the similarity between the vertices. Karypis and Kumar (1998) use METIS for final partitioning.

Hyper-graph partitioning in HGPA is done by cutting the minimal edges. Each cluster from input clustering represents the hyper-edge. Initially, the same weight is assigned to all the hyper-edges. The hyper-edge is chosen by the algorithm such that the hyper-graph is separated into K-components. The initial cluster is approximately similar to the size of K and uses the HMETIS package.

In MCLA, meta-clusters are formed which hold the clustering of clusters. The object is assigned with object-wise weight assignment which provides the cluster membership. From the graph, the hyper-edges are grouped and each data object is assigned to a meta-cluster in which its participation is the strongest.

The model of fuzzy theory is incorporated into consensus clustering framework for improving the final results. The consensus clustering approach based on fuzzy C-means is explored by Punera and Ghosh (2018). The underlying structure of various datasets was discovered by Mok, Huang, Kwok, and Au (2012) based on an ensemble framework using a fuzzy C-means algorithm. A fuzzy consensus function for ensemble clustering was studied by Sevillano, Alías, and Socoró (2012). The biological interpretation of clusters was provided by Avogadri and Valentini (2009) using a random projection technique based on a fuzzy clustering ensemble framework. A hybrid fuzzy cluster ensemble framework was proposed by Yu, Chen, You, Han, and Li (2013). In this framework, a set of associated fuzzy membership functions, the fuzzy C-means algorithm, and the affinity propagation algorithm are integrated.

6.3 DATASET DESCRIPTIONS

The dataset is a collection of pollutants for eight places, namely Tirupathi, Vishakapatnam, Gaya, Velachery, Ahmedabad, BTM Layout, Nagpur, and Agra in India. The data is taken from the Central Pollution Control Board (CPCB) (2019), which records data every 15 minutes from among various air stations located in different states in India. For a particular place and particular year, the daily values of these pollutant values have been recorded. The recorded data in the CPCB website consist of various parameters among which the air pollutant parameters are chosen and air quality index is identified. Details about the pollutant parameters are given in Section 6.4.

To add to the control measures in place, the administration of India has also implemented various sources to measure and record air pollution data. The sources include: the Centre for Geographic Analysis (CGA), which is maintained by Harvard University; the Central Pollution Control Board (Ministry of Environment, India); the Central Pollution Control Board Ministry of Environment (Historical Ambient AQI); the World AQI Project; the Delhi Pollution Control Committee, and the EPoD India-Air Pollution and Health Project.

6.4 METHODOLOGY

The data was consolidated for the years 2016–2018. The pollutant parameters included are $PM_{2.5}$, SO_2, CO, and O_3. The pre-processing involves handling missing values and identifying outliers:

- Through scatter plotting, the values having a high range of above 500 were found to be outliers.
- Missing values in the data are imputed with the overall mean value.

Year-wise data is consolidated and three clustering algorithms, namely hierarchical clustering, fuzzy C means clustering, and K-means, are applied. For each set of data, the optimal number of clusters was determined to be two, using the elbow method.

The overall procedure for weighted ensemble graph clustering is:

1. The air pollution dataset is loaded.
2. Data has to be pre-processed before applying the clustering techniques.
3. The first pre-processing step concerns the values for the chosen Air Quality Index (AQI) parameters ($PM_{2.5}$, SO_2, O_3, and CO) which are taken and loaded in a data frame.
4. The outliers are identified and replaced with the mean value.
5. Similarly, the missing values in the data will be imputed with the mean value.
6. In the final pre-processing step, the data for the various regions in India are consolidated, based on the year.
7. The pre-processed data is then used to determine the optimum number of clusters for each set of year-wise data. The number of clusters selected for each dataset is given in Figure 6.4.

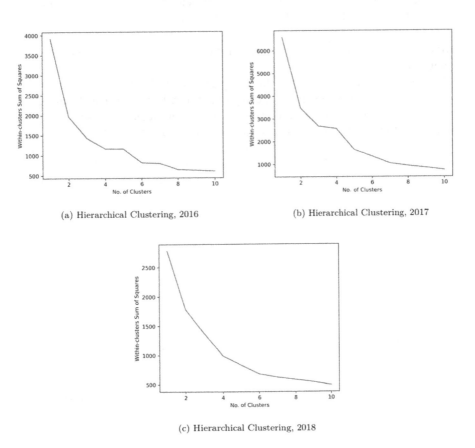

(a) Hierarchical Clustering, 2016   (b) Hierarchical Clustering, 2017

(c) Hierarchical Clustering, 2018

FIGURE 6.4   Optimal cluster plot of a base clustering algorithm.

8. As the optimal clusters are known, the base clustering techniques are applied, which are hierarchical clustering using ward linkage, fuzzy C-means clustering, and K-means clustering.
9. The results from the base models are represented in binary form using Equation 6.1 of the object membership in the respective clusters.
10. Each object weight is obtained from every pair of objects using Equations 6.2–6.5.
11. The Jaccard coefficient is calculated as the proportion of the intersection of objects to the union of objects in different clusters. Cluster weightage assignment is calculated using Jaccard similarity (using Equation 6.6).
12. The final cluster weights are used to form the similarity matrix which is an input to the graph-partitioning algorithm METIS.
13. METIS gives us the list of partition indices which are used to merge and obtain the final ensemble clustering output.
14. The final cluster labels along with the original pre-processed data are used to measure the effectiveness of the ensemble clustering technique.
15. As ground truth labels are not available in the case of clustering, intrinsic evaluation metrics such as the Davies–Bouldin Index, the Calinski-Harabasz Index, and the Silhouette Coefficient are used.

The evaluation metric results for the base clustering and ensemble clustering are then compared.

From the results of each base clustering algorithm, the cluster labels are represented as a binary membership value using Equation 6.1 (Alqurashi and Wang 2019). If the object is present in the particular cluster, then the value is 1, else 0. Figure 6.4 gives the optimal number of clusters for each base clustering algorithm, as mentioned in Table 6.1. The representation is given for hierarchical clustering for the years 2016, 2017, and 2018. The graph shows the optimal number of clusters as two. In the same way, the optimal number of clusters are chosen for the fuzzy C-means and K-means algorithm. So the co-association matrix of 6×6 is produced. The sample data is given in Table 6.2.

$$\delta\left(x_i, c_j\right) = \begin{cases} 1, & \text{if } x_i \in c_j, \forall_i = 1...n < 0 \\ 0, & \text{if } x_i \notin c_j \end{cases} \tag{6.1}$$

TABLE 6.1

Optimal Number of Clusters for Base Clustering

	Optimal Clusters		
	K-means	Fuzzy C Means	Hierarchical Clustering
2016	2	2	2
2017	2	2	2
2018	2	2	2

TABLE 6.2

Binary Representation of Base Clustering Models

	HC_0	HC_1	FCM_0	FCM_1	KM_0	KM_1
x_1	0	1	1	0	0	1
x_2	0	1	1	0	0	1
x_3	0	1	1	0	0	1
x_4	0	1	1	0	0	1
x_5	0	1	1	0	0	1
x_6	0	1	1	0	0	1
x_7	0	1	1	0	0	1
x_8	0	1	1	0	0	1
x_9	1	0	0	1	1	0
x_{10}	1	0	1	0	1	0
—	—	—	—	—	—	—
—	—	—	—	—	—	—
—	—	—	—	—	—	—
x_{7670}	1	0	0	1	1	0
x_{7671}	1	0	0	1	1	0

Note: FCM: fuzzy C means; HC: hierarchical clustering.

In the next step, using all the base clustering results, a one-shot weight assignment is developed for the objects. The object weight assignment formula is given by:

$$A_{ij} = \frac{V_{ij}}{R} \tag{6.2}$$

where V_{ij} is the co-occurrence count of the objects x_i and x_j within the same cluster and the ensemble size is given by R and A_{ij} [0, 1]. $A_{ij} \approx 1$ means that x_i and x_j are grouped into the same cluster. $A_{ij} \approx 0$ means that x_i and x_j are grouped into separate clusters. When $A_{ij} \approx 0.5$, half of the clusterings group the objects x_i and x_j together and the other half place them in different clusters. This is an uncertain scenario, since it shows no agreement on how to cluster the objects x_i and x_j. This trend is captured in a quadratic function by mapping A_{ij}, which is given by $y(x) = x(1 - x)$, where $x \in [0, 1]$. The graph of the quadratic function is given in Figure 6.5. The level of uncertainty between the two clustering objects x_i and x_j is measured by:

$$confusion\left(x_i, x_j\right) = A_{ij}\left(1 - A_{ij}\right) \tag{6.3}$$

When $A_{ij} = 0.5$, the maximum value of 0.25 of the confusion index is reached. The value of the confusion index is 0 when the value of $A_{ij} = 1$ or $A_{ij} = 0$. Each object's weight is defined by the confusion measure using:

$$w_i' = \frac{4}{n}\sum_{j=1}^{n} confusion\left(x_i, x_j\right) \tag{6.4}$$

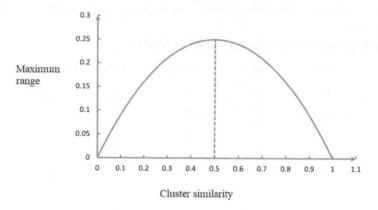

FIGURE 6.5   Quadratic function $y(x) = x(1 - x)$.

We add a smoothing term using Equation 6.5 to avoid instability by making sure that the weight value is not equal to 0.

$$w_i = \frac{w_i' + e}{1 + e} \tag{6.5}$$

where a small positive number equal to 0.01 is taken as e in our experiment. The result is $w_i \in (0, 1]$. The weight assignment of each cluster is computed by using a binary Jaccard measure given in Equation 6.6 which provides the similarity of proportion between the size of intersection and size of the union for objects within each pair of clusters C_i and C_j and associated object weights.

$$S_{ij} = \frac{\sum_{x_k \in C_i \cap C_j} w_k}{\sum_{x_k \in C_i \cup C_j} w_k} \tag{6.6}$$

From Figure 6.6, the weight between the pair of clusters is given by:

$$S_{ij} = \frac{3 \times 0.8 + 0.2 + 0.6}{4 \times (0.8 + 0.2) + 0.4 + 0.6} = \frac{3.2}{5} = 0.64$$

where the kth object's weight is represented as w_k and is calculated from Equations 6.2–6.5 and $S_{ij} \in [0, 1]$. If $S_{ij} = 0$, there is no edge between C_i and C_j in Graph G. When $S_{ij} = 1$, this indicates C_i and C_j contains the same points. S_{ij} is a similarity measure which we use to partition the meta-graph into k^* meta-clusters. The partitioning algorithm adopted is METIS, which was developed by Karypis and Kumar (1998). The output of METIS is the assignment of each object to a meta-cluster based on the ratio of participation in each cluster. Finally, the object is assigned to the cluster which has the highest ratio. The optimal number of clusters from the cluster similarity matrix is found to be three clusters. The METIS function takes this as an input. The sample output of the METIS results are explained in the following section.

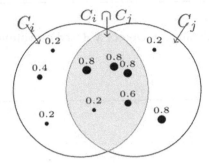

FIGURE 6.6 Similarity measure between the clusters C_i and C_j.

6.4.1 Final Cluster Labeling

The METIS function gives two outputs in the form of a tuple. The first is the objective function's value that was minimized and the second one is a list of partition indices. Our method focuses on ensemble clustering which involves assigning object weights which we use to determine cluster weights, which are converted into an adjacency list which the METIS function takes as an input. The output of the list of partition indices gives us the columns of the binary representation that have to be merged to obtain the final class labels. Once the columns are merged, their values are compared and the cluster is assigned based on the final voting value.

As an example, for representing the single object, the binary representation is given in Table 6.3. According to the METIS function, the output of [1, 2, 2, 0, 0, 1] represents the columns to be merged, which is shown in Table 6.4. Now, the merged columns give the values shown in Table 6.5. Finally, the object is assigned to the cluster that gained the majority of the votes, as shown in Table 6.6.

TABLE 6.3
Binary Representation for a Single Object along with the METIS Output

Object	HC_0	HC_1	FCM_0	FCM_1	KM_0	KM_1
X	0	1	1	0	0	1
METIS output	1	2	2	0	0	1

TABLE 6.4
METIS Output to Base Learner Output

Object	0	0	1	1	2	2
X	0	0	0	1	1	1

TABLE 6.5
Cluster Labeling for Object X Based on Voting Scheme

Object	Class Label
X	Cluster 2

TABLE 6.6
Aggregated Counts of Predictions by Base Learners

Cluster Label	Value
Cluster 0	0
Cluster 1	1
Cluster 2	2

The final clusters are generated and visualization is given in geography (Figure 6.7). The cluster results from the METIS function are partitioned into good, unhealthy, and hazardous for a given air pollution dataset. The states that fall under these categories are plotted in Figure 6.7.

6.4.2 METIS Function

In comparison to various spectral partitioning algorithms, the METIS algorithm is known to consistently produce 10–50% better results for graph partitioning (Karypis and Kumar 1998). It is a group of serial programs used to reduce the orderings for sparse matrices, to partition finite element meshes, and to partition graphs. Multi-constraint partitioning, multi-level k-way, and multi-level recursive-bisection schemes form the basis of the algorithms implemented in METIS, which is used to partition undirected graphs because the directionality of an edge does not play any role during partitioning. This is because if an edge (u, v) is cut, the edge (v, u) will also be cut. For a directed graph to be partitioned by METIS, it first needs to be converted into an undirected graph by adding an edge (v, u) for each directed edge (u, v).

6.4.2.1 METIS Algorithm
- Given the graph G = (N, E, EW)
  - Where, the nodes or vertices are represented by N;
  - The edges are represented by E;
  - The edge weights by EW.
- Where N symbolizes clusters, an edge (j, k) in E can mean that an object j in $EW_{(j,k)}$ is related to an object k.

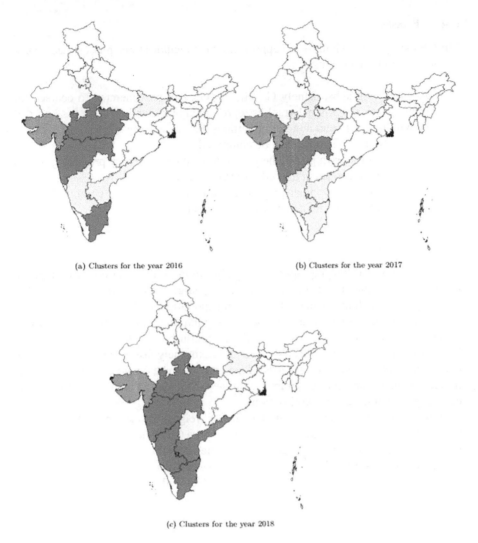

(a) Clusters for the year 2016   (b) Clusters for the year 2017

(c) Clusters for the year 2018

FIGURE 6.7 Ensemble clustering of air quality data with three clusters based on the severity of air pollution, i.e., good, unhealthy, and hazardous, represented by green, yellow, and red respectively.

- A partition $N = N_1 \cup N_2 \cup \cdots \cup N_p$ must be chosen such that
  - Load is balanced. There exists an even distribution of the weights in each N_j.
  - The parallel overhead is decreased. There should be minimization of the sum of all edge weights of edges that connect all the different partitions.
- Work will be evenly divided and communication will be minimized.
- A recursive application of partitioning the graph into two parts (graph bisection) can be carried out for complete partitioning.

6.4.3 Phases

METIS incorporates a multi-level approach which includes three phases. Each phase includes various algorithms:

- A series of graphs, namely G_0, G_1, ..., G_N, can be generated to coarsen the graph. Here, the original graph is represented by G_0 and for each $0 \leq i \leq j \leq N$, G_i will have a greater number of vertices than G_j will have.
- The partition of G_N must then be computed.
- With respect to each graph, the partition must be refined by projecting it back through the sequence in the order of $(G_N, ..., G_0)$.
- A partition of the original graph is the final partition that is computed during the third phase (the refined partition projected onto G_0).

6.4.4 Advantages

METIS is chosen as the graph-partitioning algorithm as it has various advantages. It is known to provide high quality partitions. Very large-scale integration (VLSI), linear programming, finite element methods, transportation, and various other domains in which experiments have been conducted on graphs have shown that in comparison to other widely used algorithms the partitions produced by METIS are consistently better. Another advantage of METIS is that it is extremely fast. In comparison to various other partitioning algorithms METIS has proven to be one to two orders of magnitude faster. It can partition graphs with several nodes in almost no time using current generation PCs and workstations.

Tables 6.7, 6.8, and 6.9 represent the cluster weights for each pair of clusters for the years 2016, 2017, and 2018, respectively. Figure 6.8 is the scatter plot representation of three clusters for three years based on the air pollutants. The pollutant points are represented in three different colors for three different clusters. A scatter plot is represented for the three main pollutants $PM_{2.5}$, CO, and O_3 for the year 2016, 2017, and 2018.

TABLE 6.7
Cluster Weights (S_{ij}) for the Year 2016

	C_1	C_2	C_3	C_4	C_5	C_6
C_1	1	0	0.159888	0.840112	0.909521	0.090479
C_2	0	1	0.840112	0.159888	0.090479	0.909521
C_3	0.159888	0.840112	1	0	0.069408	0.930592
C_4	0.840112	0.159888	0	1	0.930592	0.069408
C_5	0.909521	0.090479	0.069408	0.930592	1	0
C_6	0.090479	0.909521	0.930592	0.069408	0	1

TABLE 6.8
Cluster Weights (S_{ij}) for the Year 2017

	C_1	C_2	C_3	C_4	C_5	C_6
C_1	1	0	0.933622	0.066378	0.933305	0.066695
C_2	0	1	0.066378	0.933622	0.066695	0.933305
C_3	0.933622	0.066378	1	0	0.999683	0.000317
C_4	0.066378	0.933622	0	1	0.000317	0.999683
C_5	0.933305	0.066695	0.999683	0.000317	1	0
C_6	0.066695	0.933305	0.000317	0.999683	0	1

TABLE 6.9
Cluster Weights (S_{ij}) for the Year 2018

	C_1	C_2	C_3	C_4	C_5	C_6
C_1	1	0	0.793519	0.206481	0.247645	0.752355
C_2	0	1	0.206481	0.793519	0.752355	0.247645
C_3	0.793519	0.206481	1	0	0.046994	0.953006
C_4	0.206481	0.793519	0	1	0.953006	0.046994
C_5	0.247645	0.752355	0.046994	0.953006	1	0
C_6	0.752355	0.247645	0.953006	0.046994	0	1

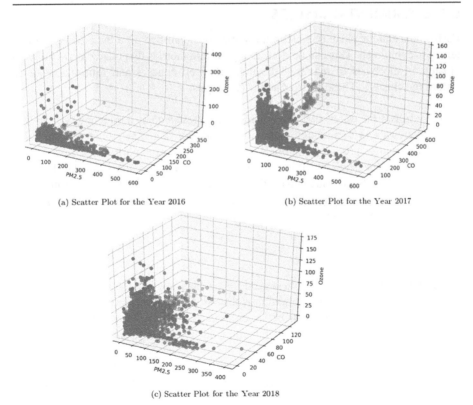

(a) Scatter Plot for the Year 2016   (b) Scatter Plot for the Year 2017

(c) Scatter Plot for the Year 2018

FIGURE 6.8   Scatter plot of ensemble clustering for AQI parameters.

Algorithm 6.1:   Weighted Ensemble Graph Clustering Algorithm

---

1: Input: K*, METIS, Base clustering Partition C
2: Output: C* which is the final Ensemble Cluster
3: Begin
4: Transfer C cluster partitions into binary vectors using Equation 6.1.
5: Using the one shot weight assignment as given in Equation 6.2, the
   objects are assigned weights.
6: Compute the cluster weight between the pair of clusters using Equation 6.6.
7: Construct a meta-graph $G = (V, E)$, in which the Equation 6.6 is used to
   compute the weights associated with the edges.
8: *MetaClusters* = METIS(G, $k*$) [*The meta-graph is partitioned into k*
   meta-clusters*]
9: $C*$ = *Assign* (*metaClusters*) [*Every object is assigned to the meta-cluster
   in which its participation is the strongest*]
10: Return $C*$
11: End

6.5 EXPERIMENTAL RESULTS

Performance of the clustering techniques must be quantified after the process of clustering is completed. This can be done using certain metrics that characterize an ideal
cluster on the basis of maximal distance between clusters and minimal distance
within the clusters. These metrics are of two types:

- Extrinsic measures such as V-measure, completeness, homogeneity, Fowlkes
  Mallows scores, mutual information based scores, and an adjusted Rand index
  that requires ground truth labels to evaluate clustering performance.
- Intrinsic measures such as the Davies-Bouldin Index, the Calinski-Harabasz
  Index, and the Silhouette Coefficient. These do not require ground truth labels
  and instead evaluate clustering performance based on the original data and the
  results of the clustering.

6.5.1 SILHOUETTE COEFFICIENT

The model with generated results is used for evaluation when the ground truth is not
known. One type of such an evaluation is the Silhouette Coefficient. It is defined for
each sample, and a model with better defined clusters is represented by a higher
Silhouette Coefficient score. The Silhouette Coefficient consists of two scores:

- Average distance between a sample and the other points within the same class.
- Average distance between a sample and the other points in the next nearest
  cluster.

TABLE 6.10

Year-Wise Silhouette Coefficient Metric for Base Clustering Model

Year	Hierarchical clustering	Fuzzy C means	K means
2016	0.657571268	0.723436528	0.684701362
2017	0.602559157	0.707689134	0.598002859
2018	0.632559157	0.658438986	0.521425682

For a single sample, the Silhouette Coefficient s is given by:

$$s(x) = b(x) - \frac{a(x)}{\max\left[b(x), a(x)\right]} \tag{6.7}$$

The coefficient value is bounded by −1 and +1; −1 indicates poor or incorrect clustering; values close to +1 indicate highly dense clustering which are well separated. Values near zero indicate overlapping clusters. The base clustering results given in Table 6.10 include the Silhouette Coefficient values for the years 2016, 2017, and 2018. In comparison to the evaluation metrics on the base clustering results, the Silhouette Coefficient for the final METIS output clusters are higher, which proves the efficiency of the ensemble cluster method. The base clustering results also show high Silhouette coefficient values of around 0.521425682 to 0.723436528. The highest Silhouette Coefficient value of the ensemble clustering technique, however, was calculated as 0.751745327.

6.5.2 Calinski-Harabasz Index

Another evaluation metric is the Calinski-Harabasz index, which is also named the variance ratio criterion. The index is the proportion of the sum of intra-cluster variation to the inter-cluster variation for all clusters (where the sum of the square of the distances is taken as the variation). The Calinski-Harabasz score s is defined as the proportion of the mean inter-cluster variation to the intra-cluster variation where E represents a set of data of size n_E which has been clustered into k clusters.

$$s = \frac{t_r(B_k)}{t_r(W_k)} \times \frac{n_E - k}{k - 1} \tag{6.8}$$

where the trace of the intra-cluster variation matrix is represented by $t_r(W_k)$ and the trace of the inter-cluster variation matrix is represented by $t_r(B_k)$. The intra-cluster and inter-cluster variation matrices are defined as:

$$W_k = \sum_{q=1}^{k} \sum_{x \in C_q} (x - c_q)(x - c_q)^T \tag{6.9}$$

TABLE 6.11

Year-Wise Calinski-Harabasz Index for Base Clustering Model

	Calinski-Harabasz Index		
Year	HC	FCM	KM
2016	1542.598346	2285.12543	1302.024934
2017	1953.759845	2016.330664	1763.262967
2018	1682.570628	1945.748552	823.3951025

$$B_k = \sum_{q=1}^{k} n_q \left(c_q - c_E\right)\left(c_q - c_E\right)^T \tag{6.10}$$

where C_q is the set of points in cluster q, c_q is the center of cluster q, c_E is the center of E, and n_q is the number of points in cluster q.

A higher value of the Calinski-Harabasz Index suggests a more optimal clustering size as it indicates the proper partition of clusters within.

As a bigger proportion of inter-cluster variance to intra-cluster variance is required to prove effective clustering, the evaluation metric results of the ensemble clustering technique give a maximum value of 2369.598641. The values provided in Table 6.13 demonstrate that the ensemble clustering technique is more effective compared to the individual base clustering techniques as the evaluation metric results for the latter range from 823.3951025 to 2285.12543. The values of the Calinski-Harabasz Index are given in Table 6.11 for all the base clustering techniques for the years 2016, 2017, and 2018.

6.5.3 DAVIES-BOULDIN INDEX

The third evaluation metric is the Davies-Bouldin Index which signifies the average similarity between the clusters by comparing the distance between them with their size (Table 6.12). A better separation between the clusters is represented by a lower Davies-Bouldin Index. The lowest possible score is zero and a better partition is indicated by values closer to zero. The index is defined as the average similarity

TABLE 6.12

Year-Wise Davies-Bouldin Index for Base Clustering Model

Year	HC	FCM	KM
2016	0.846300725	0.608585734	0.834993702
2017	0.775144086	0.565130431	0.664321735
2018	1.097000041	0.895981891	0.776500993

TABLE 6.13
Performance Matrices for Ensemble Clustering

	Ensemble Clustering		
	Silhouette Coefficient	**Calinski-Harabasz Index**	**Davies-Bouldin Index**
2016	0.713306312	2369.598641	0.55407292
2017	0.716547345	2234.848156	0.422597204
2018	0.751745327	1965.403228	0.559457489

between each cluster C_i for $i = 1, ..., k$ and its most similar one C_j. The measure R_{ij}, in the context of this index, is the similarity that trades off:

- The cluster diameter which is defined as the mean distance between the centroid of the cluster and every other point in the cluster and is taken as s_i.
- The distance between centroids i and j of the cluster and is taken as d_{ij}.

R_{ij} is non-negative and symmetric and is given by:

$$R_{ij} = \frac{s_i + s_j}{d_{ij}} \tag{6.11}$$

The Davies-Bouldin Index can be defined as:

$$DB = \frac{1}{k} \sum_{i=1}^{k} \max_{i \neq j} R_{ij} \tag{6.12}$$

The Davies-Bouldin Index values of the base clustering techniques range from 0.846300725 in hierarchical clustering, to 0.565130431 in fuzzy C-means clustering, to 0.776500993 in K-means clustering. Whereas in the ensemble clustering technique with three final clusters, the Davies-Bouldin Index values are between 0.422597204 and 0.559457489.

From these Davies-Bouldin Index values that are high, we can infer that the performance of the ensemble clustering technique exceeds that of the base clustering techniques as it has lower intra-cluster distance and greater inter-cluster distance. The final ensemble clustering results are tabulated in Table 6.13.

6.6 CONCLUSION

The focus of this chapter was to find a solution to solve the challenging task of predicting air quality based on its trends, recorded at different places at different periods of time. It is a computationally complex problem as the model needs to manage large amounts of data and to analyze patterns in order to make accurate predictions. The ensemble clustering technique was chosen as it is adapted to handle such diversified data in order to recognize patterns precisely. The first step to this approach involves preprocessing the raw dataset with the knowledge of AQI parameters such as CO, $PM_{2.5}$,

SO_2, and O_3. Results from the HC, FCM, and K-means clustering (KM) base learners are generated for ensemble learning. From these results, the object weights and cluster weights are determined. These values are then used in the METIS graph partitioning algorithm by aggregating the results from the base learners to obtain the final ensemble cluster. Higher values of the Silhouette Coefficient and the Calinski-Harabasz Index, and lower values of the Davies-Bouldin Index indicate the effectiveness of the ensemble clustering technique for the air pollution dataset that was used. Thus, the values provided for each clustering technique and the comparison of these values suggest that the weighted ensemble graph clustering algorithm provides better results as compared to the other methods. This algorithm has better cluster quality for all three measures with 0.75, 2369.59, and 0.56 indexed values for the Silhouette method, the Calinski-Harabasz Index, and the DB Index respectively. It is better than the traditional cluster quality. Visualization of the results in a Geograph of India reveals the severity of air pollution at various places for a given period. This can help one analyze the degree of pollution at various places, for future corrective measures. One of the limitations of this model is that the results do not cover every area of the country, as the air quality data is unavailable for remote areas. This limitation can be overcome by extrapolating from the clustering results for surrounding areas. In conclusion, the weighted ensemble graph clustering algorithm is a highly effective technique to forecast air quality in any chosen location.

REFERENCES

Alqurashi, T. and W. Wang. Clustering ensemble method. *International Journal of Machine Learning and Cybernetics*, 10(6):1227–1246, 2019.

Asgari, M., M. Farnaghi, and Z. Ghaemi. Predictive mapping of urban air pollution using apache spark on a hadoop cluster. In *Proceedings of the 2017 International Conference on Cloud and Big Data Computing*, pp. 89–93. ACM, New York, 2017.

Avogadri, R. and G. Valentini. Fuzzy ensemble clustering based on random projections for DNA microarray data analysis. *Artificial Intelligence in Medicine*, 45(2–3):173–183, 2009.

Ayad, H. and M. Kamel. Finding natural clusters using multi-clusterer combiner based on shared nearest neighbors. In *International Workshop on Multiple Classifier Systems*, pp. 166–175. Springer, Berlin, 2013.

Ben Ishak, A., M. Ben Daoud, and A. Trabelsi. Ozone concentration forecasting using statistical learning approaches. *Journal of Materials and Environmental Science*, 8(12):4532–4543, 2017.

Breiman, L. Bagging predictors. *Machine Learning*, 24(2):123–140, 1996.

Choi, J. Y. and B. Lee. Combining LSTM network ensemble via adaptive weighting for improved time series forecasting. *Mathematical Problems in Engineering*, Article ID 2470171, 2018. doi.org/10.1155/2018/24701712018.

CPCB. Central Pollution Control Board, *Central Control Room for Air Quality Management – All India*, 2019. https://app.cpcbccr.com/ccr/#/caaqm-dashboard-all/caaqm-landing.

Dai, C. H., Y. P. Luo, S. Luo, P. F. Qin, and H. Peng. Correlation analysis of $PM_{2.5}$ and no2 concentrations in city ambient air of changsha. In *Advanced Materials Research*, Vol. 998, pp. 1414–1418. Trans Tech Publ, 2014.

Esposito, E., S. De Vito, M. Salvato, V. Bright, R. L. Jones, and O. Popoola. Dynamic neural network architectures for on field stochastic calibration of indicative low cost air quality sensing systems. *Sensors and Actuators B: Chemical*, 231:701–713, 2016.

Fred, A. L. N. and A. K. Jain. Data clustering using evidence accumulation. In *Object Recognition Supported by User Interaction for Service Robots*, Vol. 4, pp. 276–280. IEEE, 2002.

Ghoneim, O. A., B. R. Manjunatha, et al. Forecasting of ozone concentration in smart city using deep learning. In *2017 International Conference on Advances in Computing, Communications and Informatics (ICACCI)*, pp. 1320–1326. IEEE, Piscataway, NJ, 2017.

Ghosh, J. and A. Acharya. Cluster ensembles. *Wiley Interdisciplinary Reviews: Data Mining and Knowledge Discovery*, 1(4):305–315, 2011.

Gore, R. W. and D. S. Deshpande. An approach for classification of health risks based on air quality levels. In *2017 1st International Conference on Intelligent Systems and Information Management (ICISIM)*, pp. 58–61. IEEE, Piscataway, NJ, 2017.

Greene, D., A. Tsymbal, N. Bolshakova, and P. Cunningham. Ensemble clustering in medical diagnostics. In *Proceedings. 17th IEEE Symposium on Computer-Based Medical Systems*, pp. 576–581. IEEE, Los Alamitos, CA, 2014.

Hamerly, G. and C. Elkan. Alternatives to the k-means algorithm that find better clusterings. In *Proceedings of the Eleventh International Conference on Information and Knowledge Management*, pp. 600–607. ACM, New York, 2012.

Hu, X. Integration of cluster ensemble and text summarization for gene expression analysis. In *Proceedings. Fourth IEEE Symposium on Bioinformatics and Bioengineering*, pp. 251–258. IEEE, Los Alamitos, CA, 2008.

Ishak, A. B., Daoud, M. B., and Trabelsi, A. Ozone concentration forecasting using statistical learning approaches. *Journal of Materials Environmental Science*, 8(12), 4532–4543.

Karypis, G. and V. Kumar. Multilevelk-way partitioning scheme for irregular graphs. *Journal of Parallel and Distributed Computing*, 48(1):96–129, 1998.

Kingsy, G. R., R. Manimegalai, D. M. S. Geetha, S. Rajathi, K. Usha, and B. N. Raabiathul. Air pollution analysis using enhanced k-means clustering algorithm for real time sensor data. In *2016 IEEE Region 10 Conference (TENCON)*, pp. 1945–1949. IEEE, Piscataway, NJ, 2016.

Kuncheva, L. I. and S. T. Hadjitodorov. Using diversity in cluster ensembles. In *2004 IEEE International Conference on Systems, Man and Cybernetics* (IEEE Cat. No. 04CH37583), Vol. 2, pp. 1214–1219. IEEE, Piscataway, NJ, 2014.

Li, T., C. Ding, and M. I. Jordan. Solving consensus and semi-supervised clustering problems using nonnegative matrix factorization. In *Seventh IEEE International Conference on Data Mining (ICDM 2007)*, pp. 577–582. IEEE, Los Alamitos, CA, 2007.

Martínez-España, R., A. Bueno-Crespo, I. M. Timon-Perez, J. Soto, A. Muñoz, and J. M. Cecilia. Air-pollution prediction in smart cities through machine learning methods: A case of study in Murcia, Spain. *Journal of Universal Computer Science*, 24(3):261–276, 2018.

Mok, P.-Y., H. Q. Huang, Y. L. Kwok, and J. S. Au. A robust adaptive clustering analysis method for automatic identification of clusters. *Pattern Recognition*, 45(8):3017–3033, 2012.

Nock, R. and F. Nielsen. On weighting clustering. *IEEE Transactions on Pattern Analysis and Machine Intelligence*, 28(8):1223–1235, 2006.

Punera, K. and J. Ghosh. Consensus-based ensembles of soft clusterings. *Applied Artificial Intelligence*, 22(7–8):780–810, 2018.

Schapire, R. E.. A brief introduction to boosting. In *IJCAI*, Vol. 99, pp. 1401–1406. 1999.

Sevillano, X., F. Alías, and J. C. Socoró. Positional and confidence votingbased consensus functions for fuzzy cluster ensembles. *Fuzzy Sets and Systems*, 193:1–32, 2012.

Strehl, A. and J. Ghosh. Cluster ensembles—A knowledge reuse framework for combining multiple partitions. *Journal of Machine Learning Research*, 3(3):583–617, 2012.

Topchy, A., A. K Jain, and W. Punch. A mixture model for clustering ensembles. In *Proceedings of the 2004 SIAM International Conference on Data Mining*, pp. 379–390. SIAM, Philadelphia, PA, 2014.

Vandhana, S. and J. Anuradha. Dengue prediction using hierarchical clustering methods. In *International Conference on Design Science Research in Information Systems and Technology*, pp. 157–168. Springer, Cham, 2018.

WHO. Ambient (outdoor) air quality and health, 2018. https://www.who.int/news-room/fact-sheets/detail/ambient-(outdoor)-air-quality-and-health.

Yu, Z., H. Chen, J. You, G. Han, and L. Li. Hybrid fuzzy cluster ensemble framework for tumor clustering from biomolecular data. *IEEE/ACM Transactions on Computational Biology and Bioinformatics*, 10(3):657–670, 2013.

Zeng, Y., J. Tang, J. Garcia-Frias, and G. R. Gao. An adaptive meta-clustering approach: Combining the information from different clustering results. In *Proceedings. IEEE Computer Society Bioinformatics Conference*, pp. 276–287. IEEE, Los Alamitos, CA, 2012.

Zhang, B., Hsu, M., and Dayal, U.. K-harmonic means—A spatial clustering algorithm with boosting. In *International Workshop on Temporal, Spatial, and Spatio-Temporal Data Mining*, pp. 31–45. Springer, Berlin, 2000.

Zhu, D., C. Cai, T. Yang, and X. Zhou. A machine learning approach for air quality prediction: Model regularization and optimization. *Big Data and Cognitive Computing*, 2(1):5, 2018.

7 An Intelligence-Based Health Biomarker Identification System Using Microarray Analysis

Bibhuprasad Sahu, J. Chandrakanta Badajena, Amrutanshu Panigrahi, Chinmayee Rout, and Srinivas Sethi

CONTENTS

7.1 INTRODUCTION

In biomedical experiments, researchers have to deal with large-scale databases in which the curse of dimensionality is a serious concern. We are dealing with microarray datasets during the experimental stage in which millions of genes are monitored. Different tools are adopted by many researchers; however, microarray technology is the one that is preferred most for genome expression monitoring purposes. From a literature survey it is clearly indicated that the microarray classification technique makes a noteworthy job of identifying cancer [1–4]. In microarray datasets the sample size is very small though it contains high dimension data, so it is very hard to deal with different existing methodologies which are implemented to obtain a test result, but which is not up to the mark. So we need to adopt reliable mining approaches to identify the redundant features in raw databases. For this, feature selection is needed at the preprocessing stage. We can identify the optimal set of features that can be extracted by feature selection techniques. Two basic techniques such as filter and wrapper are primarily used. In the case of the filter approach, the selection of features is done according to different constraints, such as the behavior of the feature or the content of the feature which is based on the intrinsic characteristics of the training set, where, as in a wrapper, it uses the performance of a classifier. As it uses the target classifier, the feature selection algorithm provides better performance but it is computationally expensive. Various kinds of wrapper approaches are used by many researchers, such as Tabu Search (TS), Harmony Search (HS) [8], Differential Evolution (DE) [6], Particle Swarm Optimization (PSO) [7], Genetic Algorithm (GA) [5], Artificial Bee Colony Algorithm (ABC) [9], and Ant Colony Optimization (ACO) [10].

In this research, we have suggested a binary shuffled frog-leaping algorithm with a meta-heuristic approach for feature selection for a high dimension microarray dataset. In the first stage (preprocessing stage) we extracted the top 250 genes from the original data and extracted the relevant gene subsets identified using binary shuffled frog-leaping algorithms with a meta-heuristic approach (PSO). Twenty-five subsets of genes were extracted within the range of 10–250 with 10 in each interval. The K-nearest neighbors (KNN) classifier was used to evaluate the classification accuracy performance, which was further checked using ANN and SVM.

The remainder of the chapter is organized as follows. Section 7.2 provides a literature survey of the different feature selection methods with a meta-heuristic approach. Section 7.3 describes the proposed approach using binary shuffled frog-leaping algorithms with a meta-heuristic approach for feature selection. Section 7.4 presents the detailed proposed algorithms. Section 7.5 presents the detailed experimental setup along with a simulating environment and result analysis of the proposed models. Section 7.6 concludes the work with suggested future research steps.

7.2 EXISTING KNOWLEDGE

Many researchers have used various feature selection models for dealing with huge datasets. The techniques below are preferred by most, such as TS, HS [8], DE [6], PSO [7], GA [5], ABC [9], and ACO [10]. These models are used in various application areas. But in some cases due to premature convergence they reach towards the local optima. For the above techniques, exploitation and exploration should play a major role in implementation, which may not provide better accuracy. To avoid this limitation a binary shuffled frog-leaping algorithm with a meta-heuristic approach for feature selection is used for feature selection (SFLA + META).

A modified discrete SFLA, used for the 01 knapsack problem, is demonstrated in [11]. An investigation was done with various experimental studies. For small and medium-size knapsack problems, a discrete SFLA produces remarkably better results and perhaps an alternative solution for large knapsack problems.

For the tri-objective gray environment, in [12] the author has suggested a simulated SFLA for a project selection schedule. A modified gray SFLA is proposed due to implementation of a time limit, budget constraint, and multiple objectives. The performance of the proposed algorithm is compared with NSGA-II and a multi-objective PSO for solving NP-hard problems.

To solve feature selection problems, a binary "feature selection algorithm based on bare-bones particle swarm optimization" was presented in [13]. As per the quantitative results it reveals that the proposed approach performs best average classification accuracies of 87.5%.

A new hybrid Ant Colony Optimization Algorithm for Feature Selection (ACOFS) was presented in [14]. It is claimed, as per the result analysis, that this proposed approach maintains an effective balance between exploration and exploitation of ants during searching and strengthens the global search capability of ant colony optimization for the realization of high-quality solutions in feature selection problems. Reported results of experimental testing show that ACOFS performs the remarkable ability to generate reduced-size subsets (of salient features) and improves classification accuracy.

A hybrid approach based on Particle Swarm Optimization and Support Vector Machines (PSO-SVM) with feature selection and parameter optimization to solve feature subset selection with the setting of kernel parameters is proposed by Huang et al. [15]. To minimize the computational cost, using web service technology a data mining system was implemented. The experimental results reveal that the proposed approach identifies features with high classification accuracy.

To improve the global search in an SFL algorithm, Dash et al. [16] proposed an opposition-based learning method. This algorithm improves the local search as well as increases diversity. The proposed algorithm was tested with ten different benchmark optimization functions.

In [17] the authors used SFLA as a meta-heuristic approach and applied it to many combinational problems. But this method is unable to achieve global optima and falls into local optima. So a new SFL with a Levy fight based algorithm was proposed and tested with 30 benchmark functions. Result analysis claims that it performs better.

This literature survey inspires the implementation of a binary SFLA with a meta-heuristic approach (BSFLA-META) for dimensionality reduction of a microarray dataset. We propose a novel hybrid optimization method called Binary SFLA-PSO, which introduces PSO to BSFLA by combining the fast convergence speed of PSO and the global search strategy of Binary SFLA. Frog leaping in the Binary SFLA does not impose a limit length, which helps the frog to get out of the local optimum.

After the shuffling process, the frog with the worst fitness of the whole population is substituted by the best solution searched by PSO, which aids the population to evolve more efficiently.

7.3 CLASSIFICATION MODEL

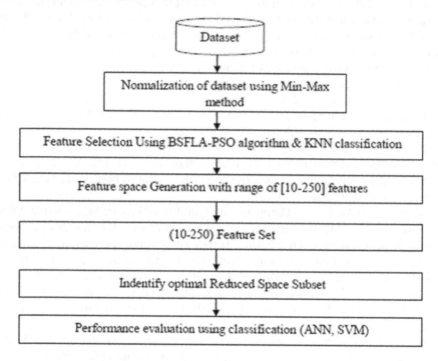

FIGURE 7.1   Proposed model.

7.4 APPROACHES FOR FEATURE SELECTION

In this section, we have implemented BSFLA-PSO. To identify 250 pre-selected genes before BSFLA-PSO, we planned for pre-feature selection.

7.4.1 SHUFFLED FROG-LEAPING ALGORITHM AND PARTICLE SWARM OPTIMIZATION (SFLA-PSO)

The Shuffled Frog-Leaping Algorithm (SFLA) is a recent evolutionary algorithm (EA) or meta-heuristic algorithm proposed by Eusuff and Lansey in 2003. It is a combination algorithm of both social behavior PSO and a genetic-based mimetic algorithm. A population of NP frogs represented the initial solution. Individual solution X_i is represented by Z dimensional vector $x_i = (x_{i1}, x_{i2}, ..., x_{ip})$ where $i \in \{1, 2, ..., NP\}$. Evaluation of the objective function of each frog is defined, and according to their fitness value frogs are arranged in decreasing order. The whole population is divided into m number of memeplexes. This is a totally circular type of activity where the first frog goes to the first memeplex; the second frog goes to the second memeplex ..., the mth frog goes to the m memeplex, and the $m + 1$th frog goes to the first memeplex [18]. While searching for food, communication between frogs occurs in two ways, either as inter-cluster or intra-cluster. Intra-cluster communication is also called local exploration, and inter-cluster communication is called global communication [19]. The frog with the best fitness is denoted x_g and the worst x_w according to their fitness value. During the evolution process the frog having the worst fitness updates its position so that it can move towards an optimal solution.

$$Z = r * \left(x_b - x_w \right) \tag{7.1}$$

$$x_{w(update)} = x_w + Z, \left\| Z \right\| \le Z_{max} \tag{7.2}$$

Here r represents a random number that varies between 0 and 1, Z_{max} is the maximum length to consider for the frog's position, and $x_{w(new)}$ gets an update with respect to the old value of $x_{w(old)}$. If there is no improvement then x_w remains constant and is replaced by randomly generated solutions. Each memeplex follows L_{max} number of iterations after the shuffling process is carried out.

Kennedy and Eberhart [20] in 1995 proposed a population-based optimization algorithm. Each population consists of NP particles called a swarm. The position value and velocity value of the ith position are represented in dimension Z, which determines the search space. The position value is represented as $x_i = (x_{i1}, x_{i2}, ..., x_{iz})$ and the velocity value represented by $v_i = (v_{i1}, v_{i2}, ..., v_{iz})$. The objective function is the main criterion for evaluating the performance of each particle. The particle's previously visited position is represented as $p_i = (p_{i1}, p_{i2}, ..., p_{iz})$ and the best previous position of each individual swarm is represented as $p_g = (p_{g1}, p_{g2}, ..., p_{gz})$. Let "itr" be the iteration number. So the swarm velocity can be represented as:

$$v_i^{itr+1} = w * v_i^{itr} + c_1 * r_1 * \left(p_i^{itr} - x_i^{itr} \right) + c_2 * r_2 * \left(p_g^{itr} - x_i^{itr} \right) \tag{7.3}$$

$$x_i^{itr+1} = x_i^{itr} + v_i^{itr+1} \tag{7.4}$$

where $i = 1, 2, 3, ..., NP$, w is the inertia weight within the range (0, 1), c_1, c_2 represents constants and learning factors, and r_1, r_2 represents a random number varying between (0, 1). Equation (7.3) represents the velocity (new) calculated from the previous velocity and the distance between the best position and current position of the particles within the population.

Equation (7.4) presents a new position if the particle changes its position in the population. The value of w (local exploration) can be computed as:

$$w = 0.4 + 0.5 * \frac{itrmax - itr}{itrmax} \tag{7.5}$$

where itr represents the current iteration number and $itrmax$ represents the number of iterations allowed.

7.4.2 THE ADVANTAGE OF SFLA

SFLA is a very simple algorithmic structure because it has less controlling parameters, but it suffers from limitations: the population searching decreases; the worst solution is updated instead of the best; and the convergence speed of the algorithm is very slow. To avoid the limitations of basic SFLA by improving frog mapping size the search learning coefficient (S) is used to change the movement of the frog during a local search. The rule of leaping of the frog is:

$$x_w \left(new \right) = x_w + stp \tag{7.6}$$

$$Stp = \varphi * Stp^{t-1} + S * rand \left(\ \right) * \left(x_b - x_w \right) - Stp_{max} \leq Stp \leq Stp_{max} \tag{7.7}$$

Stp represents step size at *t* instance of the frog step size's previous worst position iteration, φ represents the inertia required to balance the search process. The value φ decreases repeatedly from greater to smaller.

Binary shuffled frog-leaping encoding (0/1) is used for the presence and absence of the frog respectively. Each frog has two basic components such as velocity and position. The position of the frog gets updated with respect to its own position and the position of other frogs. The updated frog position is changed into a continuous value, and frogs are sorted into their fitness value. Accordingly, the whole population is divided into memeplexes and the updating of the frog in the inter-cluster occurs throughout the population. Finally, the best frog is identified.

7.4.3 ALGORITHM FOR BSFLA-PSO

For this algorithm, we considered the parameters of population size (25), the number of memeplexes (5), intra-updates of memeplexes (8), and the number of improvisations (100).

Step-1: Selection of gene expression dataset. Using Min-Max normalization normalizes the dataset. Selection of best 250 genes.

Step-2: Initialization of variables such as number of memeplexes (*m*); number of populations (*n*), improvisation with inter-cluster memeplexes (nim), number of improvisations (Ni).

Step-3: Randomly generated NP frogs as the initial population. Consider the indices either (0/1). Initialization iteration as 0.

Step-4: Using KNN classifier, calculate fitness function F_i. The frog with the highest fitness will be in the first memeplex, and so on.

Step-5: Set $i = 1$ and start a local search within inter-cluster.

Step-5.1: Set $j = 1$ and start each memeplex. Find the best and worst fitness of the whole population represented as F_{gb} and F_{gw}.

Step-5.2: Comparisons of frog fitness value between new and current. If new is greater than current then go to 8 or identify the position of the worst frog.

Step-5.3: Apply wPSO with iteration number set as $20 * L_{max}$ and find the best fitness (X_{gPSO}), the worst frog in population is replaced by X_{gPSO}.

Step-5.4: Increment j to 1 and repeat until $j = $ nim. Continue this until $i = m$.

Step-6: Check if the convergence criteria are satisfied or not with the population.

Step-7: Parameter S and φ value calculated as follows

$$S(t) = S_{low} + \left(\left(S_{high} - S_{low}\right) * t\right) / \left(\text{total no of iterations}\right)$$

$$\varphi(t) = \varphi_{low} + \left(\left(\varphi_{high} - \varphi S_{low}\right) * t\right) / \left(\text{total no of iterations}\right)$$

Step-8: Repeat steps 4–7 until iteration value not equal to Ni.

Step-9: Identify the best frog.

7.5 EXPERIMENTAL RESULT ANALYSIS

7.5.1 Dataset Considered for This Experiment

For this experimental analysis, we considered five gene expression datasets that are easily available from different repositories (Table 7.1). The course of dimensionality is the issue with the adopted datasets as the number of gene values is more than the number of observations.

7.5.2 Normalization

As datasets are drawn from different repositories there may be a chance that the data value may vary. To solve this issue we normalized the data using min–max normalization. With the help of the normalization, feature values are scaled within a particular range. After normalization there is less chance that the diversity features value misleads the objective function. Ten-fold cross-validation is always acceptable. Here we have divided the whole dataset into training and testing.

TABLE 7.1
Dataset Details

Database	No. of Genes	No. of Classes	Samples in Class 1	Samples in Class 2
Leukemia	7129	2	27	11
ALL/AML	7129	2	29	15
CNS	7129	2	39	21
ADCA Lung	12,533	2	15	134
Prostate	12,600	52	50	102

7.5.3 Details of Classifiers Used in This Experimental Study and Evaluation Metrics

In this experimental study, we used three different classifiers—SVM, ANN, and K-nearest neighbors to identify the feature data subsets. The performance evaluation of the different classifiers are compared in terms of the following criteria:

- True positive (True-pos);
- False-positive (Fal-pos);
- True negative (True-neg);
- False-negative (Fal-neg).

From these criteria we calculated the different performance evaluations of accuracy, sensitivity, and specificity.

$$\text{Accuracy} = \left[\frac{\text{True} - \text{pos} + \text{True} - \text{neg}}{\text{Total no of sample}} \right] * 100$$

$$\text{Sensitivity} = \left[\frac{\text{True} - \text{pos}}{\text{True} - \text{pos} + \text{False} - \text{neg}} \right] * 100$$

$$\text{Specificity} = \left[\frac{\text{True} - \text{neg}}{\text{True} - \text{neg} + \text{False} - \text{neg}} \right] * 100$$

7.5.4 Result Analysis

In this section, the performance of the proposed BSFLA-PSO is shown with different microarray datasets like Prostate, ALL/AML, Leukemia, ADAC Lung, and CNS. Performance evaluation of the proposed approach is tested with the KNN classifier. Later we compared the approach with basic BSFLA, Basic PSO, DE, SFLA-PSO, wPSO, and SFLA, as shown in Tables 7.2, 7.4, 7.6, 7.8, and 7.10. Figures 7.2, 7.5, 7.7, 7.11, and 7.14 present the output of three different classifiers (KNN, ANN, SVM) using the three performance metrics of accuracy, sensitivity, and specificity. Figures 7.4, 7.7, 7.10, 7.11, 7.13, and 7.16 present the error rate with different classifiers and with different datasets.

7.5.4.1 Performance of Proposed BSFLA-PSO with Prostate Dataset

Table 7.2 presents the performance of BSFLA-PSO with the prostate dataset. The result proves that taking 170 numbers of the dataset is optimal. Table 7.3 presents the performance analysis of the proposed approach with basic BSFLA, Basic PSO, DE, SFLA-PSO, wPSO, and SFLA which proves that BSFLA-PSO is the optimal and better solution.

TABLE 7.2

Classification Performance of BSFLA-PSO with the Prostate Dataset

No. of Genes	Accuracy	Sensitivity	Specificity
10	96.76	100	97.12
20	95.43	99.87	97.45
30	96.34	99.74	95.43
40	97.41	99.61	96.34
50	96.89	99.48	97.25
60	95.77	99.35	98.16
70	96.54	100	99.07
80	96.91	99.09	99.98
90	96.34	98.96	96.34
100	96.45	98.83	96.45
110	94.67	98.7	94.67
120	98.63	100	98.63
130	86.56	98.44	86.56
140	96.12	98.31	96.12
150	98.41	98.18	98.41
160	97.34	98.05	97.34
170	100	97.92	96.05
180	98.67	97.79	96.16
190	99.42	97.66	96.27
200	97.43	100	96.38
210	97.88	97.4	96.48
220	96.49	97.27	96.59
230	96.78	97.14	96.70
240	98.03	97.01	96.81
250	99.12	98.11	96.92

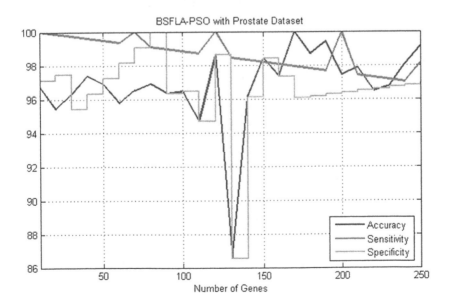

FIGURE 7.2   Comparison graph of BSFLO-PSO with prostate dataset.

TABLE 7.3

Comparison of BSFLA-PSO with Basic BSFLA, Basic PSO, DE, SFLA-PSO, wPSO, and SFLA with the Prostate Dataset

Feature Selection Approach	KNN			ANN			SVM		
	Accuracy	Sensitivity	Specificity	Accuracy	Sensitivity	Specificity	Accuracy	Sensitivity	Specificity
BSFLA-PSO	100	100	100	100	100	100	100	100	100
BSFLA	100	100	100	100	100	100	100	100	100
Basic PSO	100	100	100	100	100	100	100	100	100
DE	100	100	100	99.67	100	99.56	100	100	100
SFLA-PSO	100	100	100	100	100	100	100	100	99.68
wPSO	96.94	100	100	100	97.57	100	100	100	100
SFLA	100	100	100	100	100	100	100	100	100

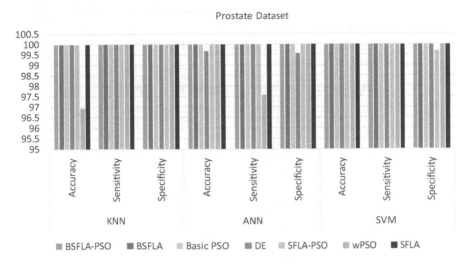

FIGURE 7.3 Performance comparison with different classifiers with prostate dataset.

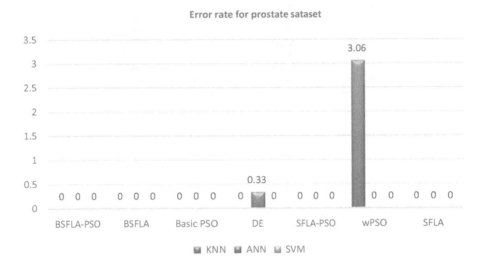

FIGURE 7.4 Error rate with prostate dataset.

7.5.4.2 Performance of Proposed BSFLA-PSO with Leukemia Dataset

Table 7.4 presents the performance of BSFLA-PSO with the leukemia dataset. The result proves that taking 130 numbers of the dataset is optimal. Table 7.5 presents the performance analysis of the proposed approach with basic BSFLA, Basic PSO, DE, SFLA-PSO, wPSO, and SFLA, which proves that BSFLA-PSO is the optimal and better solution.

TABLE 7.4

Classification Performance of BSFLA-PSO with Leukemia Dataset

No. of Genes	Accuracy	Sensitivity	Specificity
10	95.78	94.92	100.00
20	94.89	97.56	100.00
30	97.45	98.14	99.45
40	96.23	98.57	100.00
50	97.07	96.89	99.67
60	97.46	96.78	99.37
70	97.85	99.45	100.00
80	98.24	98.22	100.00
90	98.63	98.30	99.89
100	99.02	98.38	100.00
110	99.41	98.46	99.89
120	99.80	98.55	99.90
130	100.00	98.63	99.91
140	96.02	98.71	99.92
150	97.56	98.80	99.93
160	99.10	98.88	99.94
170	93.45	98.96	99.96
180	95.23	99.05	99.97
190	98.45	99.13	99.98
200	98.23	99.21	99.99
210	98.67	99.29	100.00
220	97.12	99.38	100.00
230	96.90	99.46	100.00
240	96.34	99.54	100.00
250	95.79	99.63	100.00

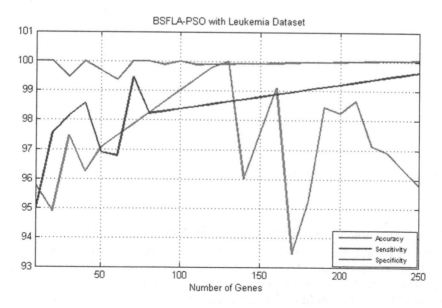

FIGURE 7.5   Comparison graph of BSFLO-PSO with leukemia dataset.

TABLE 7.5

Comparison of BSFLA-PSO with Basic BSFLA, Basic PSO, DE, SFLA-PSO, wPSO, and SFLA with Leukemia Dataset

Feature Selection Approach	KNN			ANN			SVM		
	Accuracy	Sensitivity	Specificity	Accuracy	Sensitivity	Specificity	Accuracy	Sensitivity	Specificity
BSFLA-PSO	100	100	100	100	100	100	100	100	100
BSFLA	100	100	99.33	99.57	100	100	97.58	96.79	100
Basic PSO	96.94	100	99.67	100	99.79	100	98.65	99.67	97.23
DE	96.95	100	99.88	99.46	100	100	97.89	98.57	97.68
SFLA-PSO	97.67	99.45	98.57	99.68	99.79	99.79	98.67	97.78	98.73
wPSO	95.24	98.34	98.57	97.89	97.89	98.77	95.99	95.89	98.67
SFLA	97.89	97.34	99.33	98.67	99.45	97.95	95.63	98.41	97.93

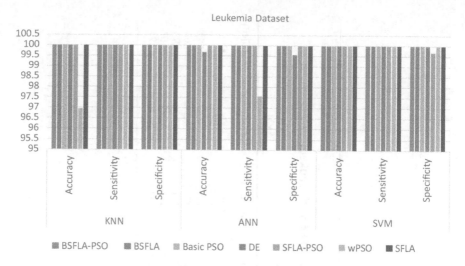

FIGURE 7.6   Performance comparison with different classifiers with leukemia dataset.

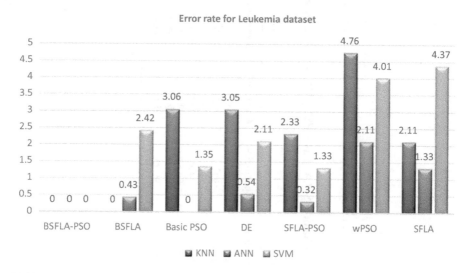

FIGURE 7.7   Error rate with prostate dataset.

7.5.4.3 Performance of Proposed BSFLA-PSO with ALL/AML Dataset

Table 7.6 presents the performance of BSFLA-PSO with the ALL/AML Dataset. The result generated proves that taking ten numbers of the datasets is optimal. Table 7.7 presents the performance analysis of the proposed approach with basic BSFLA, Basic PSO, DE, SFLA-PSO, wPSO, and SFLA, which proves that BSFLA-PSO is the optimal and better solution.

TABLE 7.6
Classification Performance of BSFLA-PSO with ALL/AML Dataset

No. of Genes	Accuracy	Sensitivity	Specificity
10	99.05	99.78	98.46
20	98.34	97.23	99.67
30	99.9	98.77	99.37
40	97.66	99.44	100.00
50	99.45	100	100.00
60	98.22	97.37	99.89
70	98.30	97.75	100.00
80	98.38	96.45	99.89
90	98.46	97.39	99.90
100	98.55	97.56	99.91
110	98.63	94.56	99.92
120	98.71	97.45	99.93
130	98.80	98.67	99.94
140	98.88	98.56	99.96
150	98.96	97.9	99.99
160	98.22	97.37	99.89
170	96.91	99.54	98.55
180	96.34	98.61	98.63
190	96.45	99.74	98.79
200	94.67	99.64	98.85
210	98.63	97.78	98.84
220	86.56	99.56	99.83
230	96.12	97.34	97.89
240	98.41	98.66	97.99
250	97.34	95.56	98.09

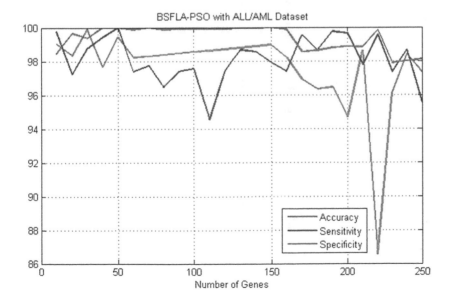

FIGURE 7.8 Comparison graph of BSFLO-PSO with ALL/AML Dataset.

TABLE 7.7

Comparison of BSFLA-PSO with Basic BSFLA, Basic PSO, DE, SFLA-PSO, wPSO, and SFLA with ALL/AML Dataset

Feature Selection Approach	KNN			ANN			SVM		
	Accuracy	Sensitivity	Specificity	Accuracy	Sensitivity	Specificity	Accuracy	Sensitivity	Specificity
BSFLA-PSO	100	100	100	100	100	100	100	100	100
BSFLA	100	100	100	100	100	100	100	100	100
Basic PSO	97.67	99.57	98.68	98.56	100	97.37	95.54	99.67	99.14
DE	98.56	97.57	97.26	96.56	96.68	98.56	100	100	100
SFLA-PSO	99.35	98.67	98.67	97.57	97.56	99.36	98.45	99.67	99.7
wPSO	98.14	99.78	98.66	96.89	96.92	97.78	97.78	99.45	99.45
SFLA	99.77	99.45	97.67	97.79	97.98	98.35	97.46	99.78	99.48

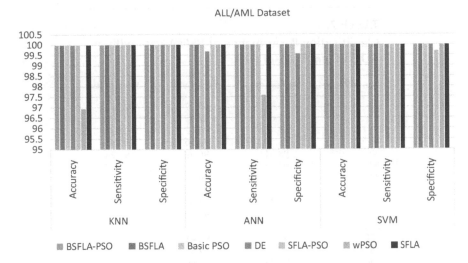

FIGURE 7.9   Performance comparison with different classifiers with ALL/AML Dataset.

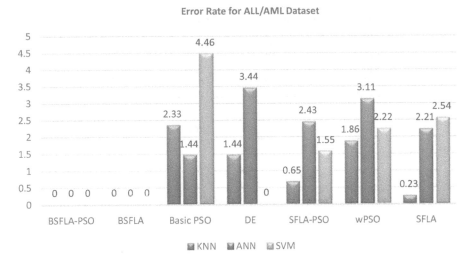

FIGURE 7.10   Error rate different classifiers with ALL/AML dataset.

7.5.4.4 Performance of Proposed BSFLA-PSO with ADCA Lung Dataset

Table 7.8 presents the performance of BSFLA-PSO with the ADCA lung dataset. The result proves that taking 100 numbers of the dataset is optimal. Table 7.9 presents the performance analysis of the proposed approach with basic BSFLA, Basic PSO, DE, SFLA-PSO, wPSO, and SFLA, which proves that BSFLA-PSO is the optimal and better solution.

TABLE 7.8

Classification Performance of BSFLA-PSO with ADCA Lung Dataset

No. of Genes	Accuracy	Sensitivity	Specificity
10	97.89	97.27	96.93
20	97.66	97.45	97.29
30	98.45	96.92	96.34
40	98.56	96.86	97.95
50	98.84	96.99	98.26
60	99.12	97.41	98.48
70	99.4	96.45	97.23
80	99.68	94.67	96.49
90	97.32	98.63	97.78
100	**99.95**	**98.99**	**99.56**
110	98.79	96.12	98.45
120	97.56	98.41	98.55
130	98.68	97.88	97.57
140	97.23	95.89	98.67
150	98.68	98.56	96.89
160	95.87	97.23	97.54
170	98.88	97.77	98.14
180	97.67	86.56	98.57
190	97.88	97.12	98.99
200	97.32	96.7	96.78
210	97.76	97.43	98.44
220	97.99	96.64	97.53
230	98.37	98.11	97.14
240	99.67	96.99	98.33
250	98.69	96.33	97.11

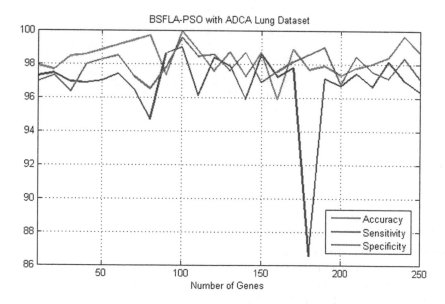

FIGURE 7.11　Comparison graph of BSFLO-PSO with ADCA lung dataset.

TABLE 7.9

Comparison of BSFLA-PSO with Basic BSFLA, Basic PSO, DE, SFLA-PSO, wPSO, and SFLA with ADCA Lung Dataset

Feature Selection Approach	KNN			ANN			SVM		
	Accuracy	Sensitivity	Specificity	Accuracy	Sensitivity	Specificity	Accuracy	Sensitivity	Specificity
BSFLA-PSO	100	100	100	100	100	99.66	100	100	100
BSFLA	97.66	99.67	99.33	99.57	100	100	97.58	96.79	100
Basic PSO	99.78	99.23	99.67	100	99.79	100	98.65	99.67	97.23
DE	98.67	98.77	99.88	99.46	100	100	97.89	98.57	97.68
SFLA-PSO	98.94	99.24	99.57	98.68	99.79	99.79	98.67	97.78	98.73
wPSO	99.78	97.02	98.57	97.89	97.89	98.77	95.99	95.89	98.67
SFLA	99.88	99.67	99.33	98.67	99.45	97.95	95.63	98.41	97.93

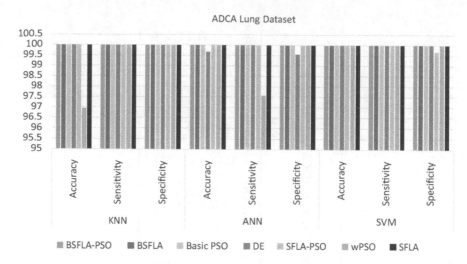

FIGURE 7.12   Performance comparison with different classifiers with ADCA lung dataset.

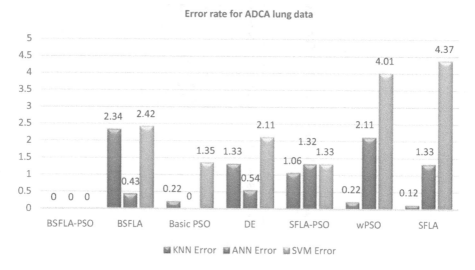

FIGURE 7.13   Error rate for ADCA lung dataset.

7.5.4.5 Performance of Proposed BSFLA-PSO with CNS Dataset

Table 7.10 presents the performance of BSFLA-PSO with the CNS dataset. The result shows that taking 90 numbers of the dataset is optimal. Table 7.11 presents the performance analysis of the proposed approach with basic BSFLA, Basic PSO, DE, SFLA-PSO, wPSO, and SFLA, which proves that BSFLA-PSO is the optimal and better solution.

TABLE 7.10

Classification Performance of BSFLA-PSO with CNS Dataset

No. of Genes	Accuracy	Sensitivity	Specificity
10	83.21	87.68	84.56
20	82.12	85.99	87.54
30	86.41	87.45	85.89
40	81.42	86.49	86.98
50	83.11	85.78	87.01
60	83.79	85.99	83.43
70	84.59	78.89	87.12
80	82.89	77.81	86.23
90	88.98	78.99	82.67
100	81.78	76.84	81.45
110	83.56	82.22	81.34
120	82.67	84.12	82.12
130	81.45	83.16	86.41
140	81.34	83.56	81.42
150	86.78	84.89	84.56
160	86.49	85.84	87.97
170	84.78	84.56	85.99
180	85.89	85.79	78.89
190	86.98	84.34	77.81
200	87.01	81.78	78.99
210	83.43	83.65	76.84
220	87.12	81.98	83.16
230	86.23	77.82	83.56
240	81.69	79.61	84.89
250	86.49	81.55	85.84

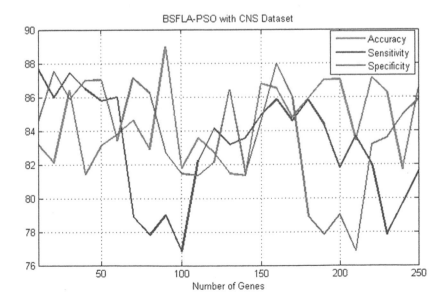

FIGURE 7.14   Comparison graph of BSFLO-PSO with CNS dataset.

TABLE 7.11

Comparison of BSFLA-PSO with Basic BSFLA, Basic PSO, DE, SFLA-PSO, wPSO, and SFLA with CNS Dataset

Feature Selection Approach	KNN			ANN			SVM		
	Accuracy	Sensitivity	Specificity	Accuracy	Sensitivity	Specificity	Accuracy	Sensitivity	Specificity
BSFLA-PSO	91.89	87.19	89.11	90.45	84.53	91.67	89.67	85.11	89.56
BSFLA	88.89	85.33	87.11	88.33	81.34	87.46	87.78	78.82	83.79
Basic PSO	87.82	82.32	83.51	86.11	83.39	86.11	87.71	78.72	83.35
DE	86.89	77.82	81.83	86.13	83.75	86.13	86.13	83.79	86.97
SFLA-PSO	88.57	85.91	78.67	85.11	82.67	86.01	87.76	83.88	87.11
wPSO	84.95	81.41	81.89	78.31	80.46	85.58	87.67	82.93	78.94
SFLA	85.91	79.45	79.91	78.56	81.66	85.78	79.67	83.67	81.22

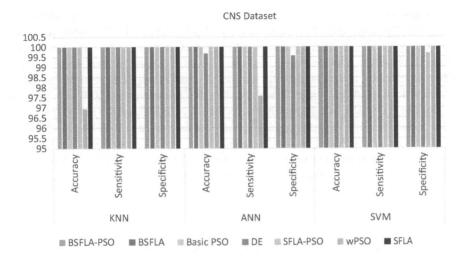

FIGURE 7.15   Performance comparison with different classifiers with CNS dataset.

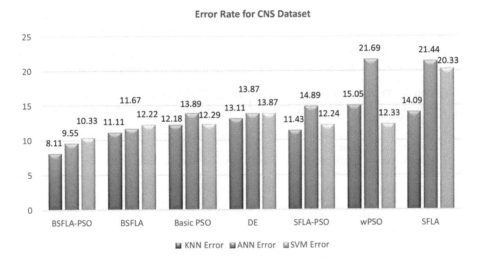

FIGURE 7.16   Comparison graph of error rate with different classifiers with CNS dataset.

7.6 CONCLUSION

In this chapter, BSFLA-PSO is a novel hybrid algorithm based on a combination of a binary shuffled frog-leaping algorithm and PSO. In the primary stage, the best 250 genes were identified and after that BSFLA-PSO was implemented; classification was done with KNN. If we deeply analyze the performance, we find that BSFLA-PSO provides outstanding performance with other different learning approaches, namely BSFLA, Basic PSO, DE, SFLA-PSO, wPSO, and SFLA. In the prostate dataset, feature selection approaches such as BSFLA-PSO, BSFLA, Basic PSO, and

SLFA perform with 100% accuracy, DE performs 100% with KNN, and SVM performs 99.67% with ANN. In the case of the leukemia dataset, BSFLA-PSO performs at 100% with all classifiers, whereas BSFLA performs at 100% with KNN only. Other approaches perform much less well. With ALL/AML and BSFLA-PSO, BSFLA performs at 100% with all classifiers, whereas only DE performs at 100% with SVM. In the ADCA lung dataset, BSFLA-PSO performs at 100% with all classifiers, but with the CNS dataset the performance of BSFLA-PSO is better than all other feature selection approaches and achieves an accuracy of 91.89%, 90.45%, and 89.67% with KNN, ANN, and SVM, respectively. The datasets considered for this experiment were binary classes in nature, so in the future we will extend this work to multiclass high-dimensional classification problems.

REFERENCES

1. Kira, K., & Rendell, L. A. (1992). A practical approach to feature selection. In *Machine learning proceedings 1992* (pp. 249–256). New York: Morgan Kaufmann.
2. Sahu, B. (2018). A combo feature selection method (filter + wrapper) for microarray gene classification. *International Journal of Pure and Applied Mathematics*, 118(16), 389–401.
3. Sahu, B., Mohanty, S. N., & Rout, S. K. (2019). A hybrid approach for breast cancer classification and diagnosis. *Endorsed Transactions on Scalable Information Systems*, 6(20), e2.
4. Sahu, B., Mohanty, S. N., & Rout, S. K. (2019). Ensemble comparative study for diagnosis of breast cancer datasets. *International Journal of Engineering & Technology*, 7(4), 15.
5. Deb, K., Pratap, A., Agarwal, S., & Meyarivan, T. (2002). A fast and elitist multi-objective genetic algorithm: NSGA-II. *IEEE Transactions on Evolutionary Computation*, 6(2), 182–197.
6. Hasan, H. B., & Kurnaz, S. (2019). Classification on breast cancer using genetic algorithm trained neural network. *IJCSMC*, 8(3), 223–229.
7. Boucheham, A., &Batouche, M. (2019). Hybrid wrapper/filter gene selection using an ensemble of classifiers and PSO algorithm. In *Biotechnology: Concepts, methodologies, tools, and applications* (pp. 525–541). Hershey, PA: IGI Global.
8. Elhoseny, M., Bian, G. B., Lakshmanaprabu, S. K., Shankar, K., Singh, A. K., & Wu, W. (2019). Effective features to classify ovarian cancer data on the internet of medical things. *Computer Networks*, 159, 147–156.
9. Sharma, R., & Kumar, R. (2019). A novel approach for the classification of leukemia using artificial bee colony optimization technique and back-propagation neural networks. In *Proceedings of 2nd international conference on communication, computing and networking* (pp. 685–694). Singapore: Springer.
10. Zainuddin, S., Nhita, F., &Wisesty, U. N. (2019, March). Classification of gene expressions of lung cancer and colon tumor using Adaptive-Network-Based Fuzzy Inference System (ANFIS) with Ant Colony Optimization (ACO) as the feature selection. *Journal of Physics: Conference Series*, 1192(1), 012019.
11. Bhattacharjee, K. K., &Sarmah, S. P. (2014). Shuffled frog leaping algorithm and its application to 0/1 knapsack problem. *Applied Soft Computing*, 19, 252–263.
12. Amirian, H., & Sahraeian, R. (2017). Solving a grey project selection scheduling using a simulated shuffled frog leaping algorithm. *Computers & Industrial Engineering*, 107, 141–149.
13. Zhang, Y., Gong, D., Hu, Y., and Zhang, W. (2015). Feature selection algorithm based on bare-bones particle swarm optimization. *Neurocomputing*, 148, 150–157.

14. Kabir, M. M., Shahjahan, M., & Murase, K. (2012). A new hybrid ant colony optimization algorithm for feature selection. *Expert Systems with Applications*, 39, 3747–3763.
15. Huang, C. L., & Dun, J. F. (2008). A distributed PSO-SVM hybrid systemwith feature selection and parameter optimization. *Applied Soft Computing*, 8, 1381–1391.
16. Dash, R., Dash, R., & Rautray, R. (2019). An evolutionary framework based microarray gene selection and classification approach using binary shuffled frog leaping algorithm. *Journal of King Saud University—Computer and Information Sciences*.
17. Shahbeig, S., Rahideh, A., Helfroush, M. S., & Kazemi, K. (2018). An efficient search algorithm for biomarker selection from RNA-seq prostate cancer data. *Journal of Intelligent & Fuzzy Systems*, 35(1), 1–10.
18. Sharma, T. K., & Pant, M. (2018). Opposition-based learning embedded shuffled frog-leaping algorithm. In *Soft computing: Theories and applications* (pp. 853–861). Singapore: Springer.
19. Sahu, B. (2019). Multi-tier hybrid feature selection by combining filter and wrapper for subset feature selection in cancer classification. *Indian Journal of Science and Technology*, 12(3), 1–11. doi:10.17485/ijst/2019/v12i3/141010
20. Kennedy, J., & Eberhart, R. (1995). Particle swarm optimization. In *Proceedings of ICNN'95-International Conference on Neural Networks* (Vol. 4, pp. 1942–1948). IEEE.

8 Extraction of Medical Entities Using a Matrix-Based Pattern-Matching Method

Ruchi Patel and Sanjay Tanwani

CONTENTS

8.1 INTRODUCTION

A massive number of research papers on disease treatment, prevention, and diagnostics have been published. The medical text data provide the origin of information for biomedical study and research. However, these research papers are scattered across a huge medical informatics literature which have been published by specialist doctors. It is difficult for doctors to read all of these publications and discover new knowledge. The need is to accumulate all the information in a single place so that the specialist doctor may obtain guiding information for the most effective treatment and prevention. Health care professionals keep patient details, such as their past medical history, signs and symptoms of diseases, tests and treatments, and medication, in clinical records like discharge summaries and patients' prescriptions. These clinical records are in the form of unstructured or semi-structured texts. Extracting medical knowledge from an unstructured clinical dataset is a real challenge.

163

The Center of Biomedical Computing, named i2b2 (Informatics for Integrating Biology and the Bedside), has organized different challenges in natural-language processing (NLP) for extracting useful knowledge from clinical texts. One of the challenges was organized in 2010, in which one task focused on extracting clinical concepts such as problems, tests, and treatments from clinical records [1]. Medical concept detection is also called a clinical entity recognition. A named entity recognition is an essential part of clinical NLP, because it is a crucial step for extracting knowledge from semi-structured or unstructured texts. Named entity recognition has two phases: entity boundary detection and entity type classification. Previously, many systems had been developed for recognizing clinical entities using a machine learning approach [1], a rule based approach [2], a dictionary look-up method [3], and a hybrid method [4]. Entity boundary detection is a type of sequence labeling problem, resolved by a BIO model, where B refers to "beginning," I to "inside," and O to "outside." Other models are IOBW and IOBEW, where E refers to "end" and W refers to "single word" [5]. Some chunkers like openNLP Chunker, Peregrine [6], Tree Tagger, and MetaMap [7] are available but do not perform precisely. MetaMap is an information extraction tool for medical texts which also uses the chunking method, but its precision and recall are both worse.

A single word entity is easy to find and classify, but boundary detection of the sequence of entities is still a vital issue in clinical text processing. For example, "oxycodone–acetaminophen," "saphaneous vein graft → posterior descending artery," and "a permanent dual chamber rate responsive pacemaker." In the proposed work the problem of correct boundary identification of clinical concepts is explored. The proposed system uses Part of Speech (POS) as a feature for training the model. The system is based on a matrix model and performs multi-pattern matching. These matched patterns are converted to their corresponding words and mapped with a unified modeling language system (UMLS) for entity classification. The rest of this chapter is organized as follows. Section 8.2 illustrates the background of clinical named-entity recognition, Section 8.3 describes the proposed method and dataset, Section 8.4 presents system evaluation, Section 8.5 provides the experimental results and a discussion, and Section 8.6 concludes and indicates some new directions for further research.

8.2 BACKGROUND

Many different NLP challenges had organized and focused on medical concept extraction tasks such as the i2b2 2010 challenge for clinical notes [8] and the ShARe/CLEF eHealth Shared Task [9], the shared task of BioNLP/NLPBA 2004 with the GENIA dataset for identifying different biomedical entities [10], and the BioCreAtIvE challenge [11] for recognizing biological concepts like gene mention identification [12].

In previous works of clinical entity identification, different methods have been used such as the dictionary look-up method, and rules-based and machine learning. The dictionary look-up method used in [3], in which the authors identified clinical entities using dictionaries compiled from the corpus, performed experimentation on I2b2 2010 dataset and obtained the average F score of 48% for the Beth dataset and

50% for the Partners dataset. The rule-based method is also used in [13, 14], in which some rules are created based on corpus words and word occurrences, and words are then found in the corpus and mapped to a corresponding category and provided a 42% F score. Machine-learning-based approaches like SVM (support vector machine) and CRF (conditional random field) have been used for entity boundary identification and entity classification [1, 15], and which is based on the beginning, inner, and outside (BOI) model for sequence labeling. An unsupervised approach has also been used to extract named entities from biomedical texts [16], in which authors have developed a noun phrase chunker followed by a filter based on inverse document frequency. The classification of multiword entities is carried out by using the concept of distributional semantics.

A number of systems have been developed for medical entity recognition, such as MetaMap, which is used for concept extraction [17]. It uses UMLS for medical-term identification [18], but its results are worse than other approaches. MetaMap 2013v2 [13] gave a 40% F score and MetaMap 2010 [6] gave a 21.8% F score; previous versions [7] gave an average 15.5% F score. Other systems have used entity boundary identification, like Peregrine (F score 46.8%), OpenNLP chunker (F score 70.0%), and StanfordNer (F score 76.8%) [6]. A lot of previous research work has been identified for entity boundary detection of medical data. In [5, 19], the authors used OpenNLP chunker for named entity recognition (NER) with IO, IOB, and IOBW models, and, for getting correct boundaries, post-processing is performed with boundary adjustment rules. In [7], the authors compared different methods like MetaMapPlus, a rule-based method, and obtained an F score of 52.28%; the CRF method, with an F score of 45.33%; and an SVM method, with an F score of 76.17%.

For text pre-processing, several NLP tools were used, like Lingpipe, Tree Tagger, OpenNLP, c-TAKES, Stanford parser [20], splitta, SPECIALIST, and Stanford CoreNLP. Evaluation of boundary detection of sentences using these tools is carried out in [21], where the authors discussed and identified different errors, such as the detection of sentence splitters like semicolons and colons, though these errors were separate from their context. As per their evaluation, except for c-TAKES, other tools performed worse on clinical notes than on general domain text.

Das et al. [22] proposed a neuro-fuzzy model with post-feature reduction to analyze complex biomedical data. In this paper, to identify uncertainty issues from input patterns, a class-belongingness fuzzification method is used. As the result of fuzzification, input patterns increase; to handle this issue, post-feature reduction is used which eliminates all the irrelevant data from the input set. Das et al. [23] proposed a framework for the classification of medical diseases. For dealing with the uncertainty of data a linguistic neuro-fuzzy with feature extraction (LNF-FE) model is presented in this paper. Where linguistic fuzzification is used for finding the membership value and, from that, values, the relevant data are retrieved using feature extraction. Finally, the reduced features are classified using an artificial neural network (ANN). Das et al. [24] used a particle swarm optimization (PSO) model for building a multilayer perceptron which is used for classification. The model is capable of solving linear and non-linear problems. The back propagation algorithm is used for training the network. The performance of the proposed model is compared with multilayer perceptron (MLP) and genetic algorithm (GA-MLP) also. Das et al. [25] used a hybrid neuro fuzzy and feature reduction model for data analysis. Using fuzzification, the input pattern classes are identified and the irrelevant data is removed by using feature reduction methods. An ANN model is used for classifying the filtered data. The performance of the model is tested against ten different datasets.

8.3 METHODOLOGY

8.3.1 DATASET

The Informatics for Integrating Biology and the Bedside Center (i2b2) organized a challenge in 2010, which was focused on NLP for clinical data. The dataset included patients' discharge summaries and clinical notes provided by three institutions: the Beth Medical Center, Partners HealthCare, and the Pittsburgh Medical Center. The organizers annotated manually 826 clinical notes, which provided gold-standard data for the challenge [8].

For developing the new system for entity boundary detection, the task of clinical concept extraction must be focused. In this task, 426 annotated progress notes of Partners and Beth were used where 170 progress notes were used as a training set, and the remaining 256 notes were used as a test dataset to assess the performance of the systems with reference data.

8.3.2 PROPOSED METHOD

The proposed system uses the concept of multi-pattern matching based on a matrix model [26] for clinical data. Fundamentally, a matrix is a rectangular array of numbers, symbols, or expressions arranged in rows and columns. The system performs the parallel pattern matching between two matrices in which those of the same size are added or subtracted, element by element. Every element of a matrix has a user-assigned value corresponding to the POS tag of an entity.

This system is created in two matrices, one for gold-standard training data and another for test data. The framework of the proposed system (see Figure 8.1) is divided into a few components for medical concept extraction. The working of the components is described below.

8.3.2.1 Text Pre-Processing

I2b2 clinical notes contain various parts, such as discharge date, admission date, allergies, present illness history, past medical history, social history, family history, physical examination, pertinent results, discharge medication, and discharge instructions. These clinical notes are in an unstructured and semi-structured form of text [27]. Every part encloses a little information associated with each patient and concerns various special characters, colons, semicolons, punctuation, hyphens, and so on. For medical concept identification, each sentence of each section needs pre-processing. In this system, a natural language toolkit is used for text processing. For performing tokenization some methods, like line tokenize or word tokenize, are used, followed by a post-tag method for parts of speech tagging. Pre-processing is performed on gold-standard annotations and test annotations. In gold-standard

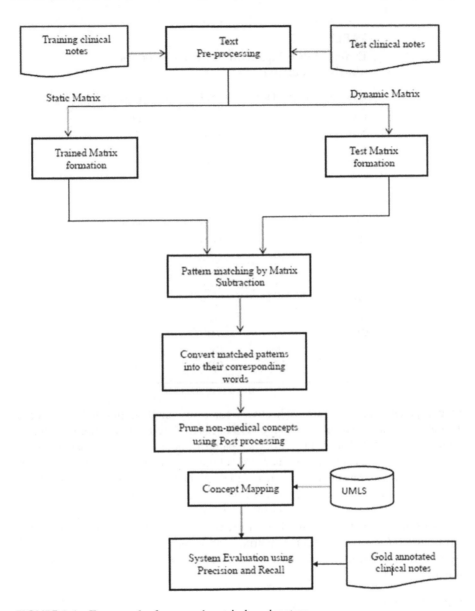

FIGURE 8.1 Framework of proposed matrix-based system.

annotations, each file has its named entities with its concepts. These entities are tagged with their POS. In a test dataset, every sentence of the file is tagged with POS. These tagged data are applied as input for creating matrices.

8.3.2.2 Trained Matrix Formation

The system is trained using gold-standard annotations for which a trained matrix was created. After the text processing of the gold-standard annotations of every file, POS

TABLE 8.1
Different POS Tags with Assigned Value for Matrix Calculation

POS Tag	Assigned Value	POS Tag	Assigned Value
NN	1	CL	18
NST	2	INTF	19
NNP	3	INJ	20
PRP	4	NEG	21
PRP$	5	UT	22
DEM	6	SYM	23
VM	7	XC	24
VAUX	8	RDP	25
JJ	9	ECH	26
RB	10	UNK	27
PSP	11	VBD	28
RP	12	VBG	29
CC	13	NNS	30
WQ	14	DT	31
QF	15	VBN	32
QC	16	IN	33
QO	17	CD	34

tagged patterns are generated collectively for every single entity and sequence of entities. For matrix calculation, every POS tag is assigned with a different value, as shown in Table 8.1. The frequency of each pattern among the different pattern sets of gold annotations is then ascertained. For obtaining frequent pattern sets, a user-defined minimum threshold value greater than or equal to 2 is required. These different frequent pattern sets are then applied as elements for the trained matrix.

The trained matrix has k number of pattern sets, where every row represents one pattern set. The longest and the shortest pattern's length are *maxl* and *minl*, respectively. In $\alpha(M \times N)$, α is a trained matrix, M is k (number of pattern sets), and N is *maxl*. When the location of the elements is lacking, "0" is entered in the columns of the matrix.

8.3.2.3 Test Matrix Formation

The test matrix (β) is created using test data for every file independently. POS tagged test data with an assigned value is applied to the matrix. Here, the number of rows and columns are the same as the trained matrix α. Every complete sentence in its converted form is passed to the matrix, which changes dynamically at the entrance of every new sentence after comparison with the trained matrix.

8.3.2.4 Pattern Matching

The system performs multi-pattern matching of the trained matrix with the test matrix by using matrix subtraction. Matrix α minus matrix β is equal to matrix θ.

If a row of θ is equal to 0 (all are 0's), this shows the pattern is completely matched. When matrices α and β are subtracted once, matrix β will transform circularly until the end of the text. Every file's test matrix is compared with the trained matrix. After obtaining exactly matched patterns, these are converted into their corresponding POS tags as are their entities or sequence of entities.

8.3.2.5 Pruning Non-Medical Concepts

The system generates several medical or non-medical single entities or sequences of entities because of the parallel pattern matching between two matrices and because the precision of non-medical concepts drops off. For improving the precision of the system, non-medical concepts are pruned according to post-processing rules. For this, a few rule patterns are shown below in which a medical concept's semantic type is one feature which is presented in the UMLS Metathesaurus [2]. UMLS shows some categories of semantic types of the database which is used for three medical concepts: problem, test, and treatment.

$$\text{Rule:}\left(\text{Semantic type} = y1\right) \vee \left(\text{Semantic type} = y2\right) \vee \left(\text{Semantic type} = y3\right) \vee$$
$$\left(\text{Semantic type} = y4\right) \vee \left(\text{Semantic type} = y5\right) \rightarrow \text{Class} = Y$$

where $Y = 1$ to m, which concerns different semantic types that are subsequent to their concept class, which is used for categorization. For example:

- *If y1 = 'Finding', y2 = 'Sign or Symptom', y3 = 'Disease or Syndrome',*
  y4 = 'Pathologic Function'
  then Y = Problem

- *If y1 = 'Laboratory or Test Result', y2 = 'Cell',*
  y3 = 'Laboratory Procedure',
  y4 = 'Tissue',
  y5 = 'Clinical Attribute' then Y = Test

- *If y1 = 'Diagnostic Procedure', y2 = 'Therapeutic or Preventive Procedure',*
  y3 = 'Organic Chemical',
  y4 = 'Pharmacologic Substance',
  y5 = 'Antibiotic'
  then Y = Treatment

If the word token is equivalent to any category of semantic type, then it can be precisely mapped to a suitable medical semantic class.

8.4 SYSTEM EVALUATION

The effectiveness of the proposed system is measured by precision and recall [28]. Here, precision is the ratio of correctly extracted medical concepts out of all extracted medical concepts. Recall is the ratio of correctly extracted medical concepts out of all medical concepts that should have been extracted. Precision and recall for the proposed system is calculated as:

$$\text{Precision} = \frac{\text{Number of patterns that are completely matched in matrix}}{\text{Number of patterns that are retrived in matrix}} \times 100 \quad (8.1)$$

$$\text{Recall} = \frac{\text{Number of patterns that are completely matched in matrix}}{\text{Number of relevant patterns that are given in gold annotation}\left(\text{trained matrix}\right)} \times 100$$

$$(8.2)$$

8.5 RESULTS AND DISCUSSION

Based on the concept of precision and recall given in Equations 8.1 and 8.2, the matrix-based proposed system generates approximately a 60% recall and a 30% precision for all concepts collectively (test, problem, treatment). The precision of the system is reduced because it uses the concept of parallel pattern matching between two matrices and generates many non-medical entities also. To improve the precision of the system, some post-processing rules are introduced, after which precision is increased to 70% approximately. A comparison of precision between before and after post-processing is given for Beth and Partners data in Figure 8.2. The performance of the system is measured for every concept (problem, test, and treatment) separately and compared with the Partners and Beth results (see Figure 8.3). It is observed that the precision of the Beth data is greater than the Partners data for the problem and treatment concepts, but for the test concept it is equal. It is found that the recall of the Partners dataset is higher than for the Beth dataset for every concept. The system is compared with different methods, such as rule based [13], dictionary look-up [3], the chunker module of MetaMap 2010 [4], and MetaMap 2013 version [13]. The precision and recall of the proposed system is better than these methods (see Figures 8.4 and 8.5).

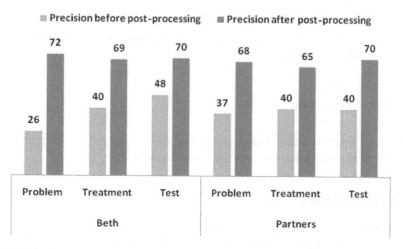

FIGURE 8.2   Comparison of precision before and after post-processing rules.

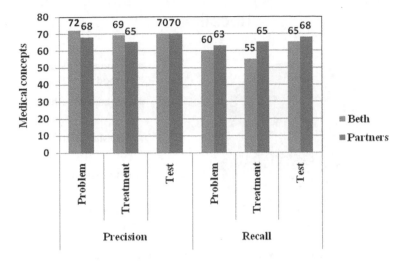

FIGURE 8.3 Comparison of performance of proposed method for treatment, problem, and test concepts based on precision and recall.

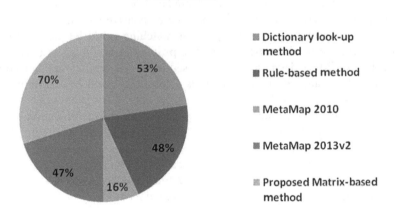

FIGURE 8.4 Comparison between precision of different text classification methods.

After conducting error analysis on the results, it is observed that improvement in recall is possible to some extent. A few concepts are missing because of the inaccurate boundary identification of sequences of entities. For frequent pattern generation, the minimum threshold value is 2 and patterns those are having frequency 2 or more than 2, they have considered. But frequency 1 patterns are missed and the corresponding words were not extracted. In this system, only the POS feature was focused. In future work, the same system could be developed by using more features, like word, n-gram, semantics of words, lemma, prefix, and suffix.

Boundary identification of multiword clinical concepts is the big issue [29] which is principally found in the problem and treatment concepts because the gold-standard

Recall

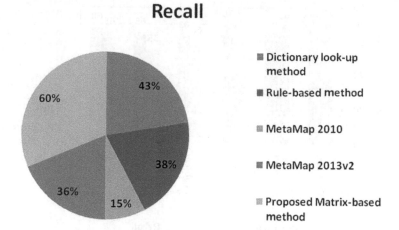

FIGURE 8.5   Comparison between recall of different text classification methods.

dataset contains a large sequence of entities in problem concepts like "minimal right internal carotid artery disease" or "thrombosis of the right renal artery." A large sequence of entities is also found in treatment concepts like "a left carotid endarterectomy," "oxycodone–acetaminophen," "saphaneous vein graft → posterior descending artery," or "a permanent dual chamber rate responsive pacemaker." The system performed strict matching; in future, partial matching could be performed which might increase the recall of the system. The pruning of non-medical concepts was done using a UMLS database, but large sequences of entities are not correctly matched in this database to medical words. This issue could be determined using relaxed matching of clinical concepts.

Text pre-processing included regular expressions for the identification of special characters, which are treated as word splitters, though only some words are identified in the gold-standard data which included these characters in between. Examples of composite words such as "severe 3 vessel disease," "saphaneous vein graft → obtuse marginal," "the patient's pacing wires," "heel/shin," and "leg, emg & apos" are inaccurately predicted as concepts using the anticipated method.

8.6 CONCLUSIONS AND FUTURE WORK

In the proposed solution, medical concepts, such as signs and symptoms of diseases, tests, and treatments, are identified using parallel pattern matching of two matrices followed by a rule-based method. The pruning of non-medical concepts increased the performance of the system by using concept mapping with semantic types of UMLS. It is noted that the efficiency of the system is improved over MetaMap 2010, MetaMap 2013v2, and dictionary look-up methods. In future, recall could be further improved by adding new features like prefix and suffix, stemming, word, semantics of words, or n-gram in the same method. Machine learning concepts are not used, which is why the performance of the system is not dependent on the size of the dataset. The proposed system could be implemented on small and large datasets in an identical way, but the processing time for large datasets would increase. In future work, for the fast processing of large text, the proposed method could be implemented in a distributed environment.

ACKNOWLEDGMENTS

We would like to thank the organizers of the 2010 i2b2 challenge for designing, training, and evaluating corpora. We also wish to thank the U.S. National Library of Medicine for allowing us to access clinical work through UMLS.

REFERENCES

1. Jiang, M., Chen, Y., Liu, M., Rosenbloom, S.T., Mani, S., Denny, J.C., Xu, H.: A study of machine-learning-based approaches to extract clinical entities and their assertions from discharge summaries. *Journal of the American Medical Informatics Association: JAMIA* 18, 601–606 (2011).
2. Kim, Y., Riloff, E.: A stacked ensemble for medical concept extraction from clinical notes. *AMIA Jt Summits on Translational Science Proceedings* 2015, 69–73 (2015).
3. Torii, M., Wagholikar, K., Liu, H.: Using machine learning for concept extraction on clinical documents from multiple data sources. *Journal of the American Medical Informatics Association: JAMIA* 18, 580–587 (2011).
4. Minard, A.-L., Ligozat, A.-L., Ben Abacha, A., Bernhard, D., Cartoni, B., Deléger, L., Grau, B., Rosset, S., Zweigenbaum, P., Grouin, C.: Hybrid methods for improving information access in clinical documents: Concept, assertion, and relation identification. *Journal of the American Medical Informatics Association* 18, 588 (2011).
5. Dehghan, A.: Boundary identification of events in clinical named entity recognition. arXiv preprint arXiv 1308, 1004 (2013).
6. Kang, N., Afzal, Z., Singh, B., van Mulligen, E.M., Kors, J.A.: Using an ensemble system to improve concept extraction from clinical records. *Journal of Biomedical Informatics* 45, 423–428 (2012).
7. Abacha, A.B., Zweigenbaum, P.: Medical entity recognition: A comparison of semantic and statistical methods. In: *Proceedings of BioNLP 2011 Workshop*, pp. 56–64. Association for Computational Linguistics, Portland, OR (2011).
8. Uzuner, Ö., South, B.R., Shen, S., DuVall, S.L.: 2010 i2b2/VA challenge on concepts, assertions, and relations in clinical text. *Journal of the American Medical Informatics Association: JAMIA* 18, 552–556 (2011).
9. Suominen, H., Salanterä, S., Velupillai, S., Chapman, W.W., Savova, G., Elhadad, N., Pradhan, S., South, B.R., Mowery, D.L., Jones, G.J.F., Leveling, J., Kelly, L., Goeuriot, L., Martinez, D., Zuccon, G.: Overview of the ShARe/CLEF eHealth evaluation lab 2013. In: Forner, P., Müller, H., Paredes, R., Rosso, P., Stein, B. (eds.), *Information Access Evaluation. Multilinguality, Multimodality, and Visualization: 4th International Conference of the CLEF Initiative, CLEF 2013, Valencia, Spain, September 23–26, 2013. Proceedings*, pp. 212–231. Springer Berlin Heidelberg, Berlin (2013).
10. Kim, J.-D., Ohta, T., Tsuruoka, Y., Tateisi, Y., Collier, N.: Introduction to the bio-entity recognition task at JNLPBA. In: *Proceedings of the International Joint Workshop on Natural Language Processing in Biomedicine and its Applications (NLPBA/BioNLP)*. Geneva, Switzerland (2004).
11. Hirschman, L., Yeh, A., Blaschke, C., Valencia, A.: Overview of BioCreAtIvE: Critical assessment of information extraction for biology. *BMC Bioinformatics* 6, S1 (2005).
12. Yeh, A., Morgan, A., Colosimo, M., Hirschman, L.: BioCreAtIvE task 1A: Gene mention finding evaluation. *BMC Bioinformatics* 6, S2 (2005).
13. Kim, Y., Riloff, E., Hurdle, J.F.: A study of concept extraction across different types of clinical notes. *AMIA Annual Symposium Proceedings* 2015, 737–746 (2015).

14. Sahu, R.: Rule-based method for automatic medical concept extraction from unstructured clinical text. In: Sa, P.K., Bakshi, S., Hatzilygeroudis, I.K., and Sahoo, M.N. (eds.), *Recent Findings in Intelligent Computing Techniques*, pp. 261–267. Springer, Berlin (2018).

15. Patel, R., Tanwani, S.: Application of machine learning techniques in clinical information extraction. In: Mishra M., Mishra B., Patel Y., and Misra R. (eds.), *Smart Techniques for a Smarter Planet*, pp. 145–165. Springer, Cham, Switzerland (2019).

16. Zhang, S., Elhadad, N.: Unsupervised biomedical named entity recognition: Experiments with clinical and biological texts. *Journal of Biomedical Informatics* 46, 1088–1098 (2013).

17. Aronson, A.R., Lang, F.-M.: An overview of MetaMap: Historical perspective and recent advances. *Journal of the American Medical Informatics Association: JAMIA* 17, 229–236 (2010).

18. Bodenreider, O.: The Unified Medical Language System (UMLS): Integrating biomedical terminology. *Nucleic Acids Research* 32, D267–D270 (2004).

19. Dehghan, A.: Boundary adjustment of events in clinical named entity recognition. *CoRR* (2013).

20. Xu, H., AbdelRahman, S., Jiang, M., Fan, J.w., Huang, Y.: An initial study of full parsing of clinical text using the Stanford Parser. In: *2011 IEEE International Conference on Bioinformatics and Biomedicine Workshops (BIBMW)*, pp. 607–614. Atlanta, GA (2011).

21. Griffis, D., Shivade, C., Fosler-Lussier, E., Lai, A.M.: A quantitative and qualitative evaluation of sentence boundary detection for the clinical domain. *AMIA Summits on Translational Science Proceedings* 2016, 88–97 (2016).

22. Das, H., Naik, B., Behera, H., Jaiswal, S., Mahato, P., Rout, M.: Biomedical data analysis using neuro-fuzzy model with post-feature reduction. *Journal of King Saud University-Computer and Information Sciences* (2020).

23. Das, H., Naik, B., Behera, H.: Medical disease analysis using neuro-fuzzy with feature extraction model for classification. *Informatics in Medicine Unlocked* 18, 100288 (2020).

24. Das, H., Jena, A.K., Nayak, J., Naik, B., Behera, H.: A novel PSO based back propagation learning-MLP (PSO-BP-MLP) for classification. In: *Computational Intelligence in Data Mining*, Volume 2, pp. 461–471. Springer, Berlin (2015).

25. Das, H., Naik, B., Behera, H.: A hybrid neuro-fuzzy and feature reduction model for classification. *Advances in Fuzzy Systems* 2020, 15 (2020).

26. Zhang, H., Xu, D., Zhang, L., Sun, Y.: Matrix-based parallel pattern matching method. In: *2015 IEEE International Conference on Communications (ICC)*, pp. 7114–7119. London, UK (2015).

27. Gong, L.J., Yuan, Y., Wei, Y.B., Sun, X.: A hybrid approach for biomedical entity name recognition. In: *2009 2nd International Conference on Biomedical Engineering and Informatics*, pp. 1–5. Tianjin, China (2009).

28. Rout, J.K., Singh, S., Jena, S.K., Bakshi, S.: Deceptive review detection using labeled and unlabeled data. *Multimedia Tools and Applications* 76, 3187–3211 (2017).

29. Rout, J.K., Choo, K.-K.R., Dash, A.K., Bakshi, S., Jena, S.K., Williams, K.L.: A model for sentiment and emotion analysis of unstructured social media text. *Electronic Commerce Research* 18, 181–199 (2018).

9 Supporting Environmental Decision Making
Application of Machine Learning Techniques to Australia's Emissions

Alex O. Acheampong and Emmanuel B. Boateng

CONTENTS

9.1 INTRODUCTION

This chapter seeks to compare the forecasting ability of machine learning (ML) techniques such as decision tree, random forest, extreme gradient boosting, and support vector regression (SVR) by focusing on Australia's carbon emissions. Australia is one of the major carbon-emitting countries in the world. The Ndevr Environmental (2019)[1] report indicates that Australia's carbon emissions for March 2019 increased to approximately 561 million tonnes of CO_2 equivalent. The continued increase in carbon emissions, which is the primary greenhouse gas behind climate change, would cause a decline in agriculture yield, damage property and infrastructure facilities, and increase commodity prices and financial instability (Climate Council, 2019). Australia has enacted numerous policies to mitigate carbon emissions; however, it is critical to have a futuristic understanding of the country's carbon emissions.

Having such an understanding requires the use of advanced modeling or forecasting techniques. In the existing literature, the majority of the studies have utilized classical statistical approaches to model or forecast carbon emissions (see Acheampong & Boateng, 2019). However, given the chaotic, non-linearity, and non-stationarity of variables for modeling carbon emissions, the classical statistical approaches are not appropriate for modeling such complex behavior (Acheampong & Boateng, 2019; Gallo, Contò, & Fiore, 2014). Apart from the classical statistical approaches, other structural simulation approaches, such as computable general equilibrium (CGE) models, the Atmospheric Stabilisation Framework (ASF) model, the Multiregional Approach for Resources and Industry Allocation (MARIA) model, and the National Energy Modelling System (NEMS), have been employed by various organizations to forecast carbon emissions (see Auffhammer, 2007; O'Neill & Desai, 2005; Zhao & Du, 2015). It is argued against these models that, since the value parameters are fixed using personal judgments and calibrations, it may not be able to capture the behavior of the real economy (cited in Zhao & Du, 2015). Further, the Grey Model (GM) has been used as a technique for forecasting carbon emissions; however, Yin & Tang (2013) argue that this model and especially GM (1,1) works well with limited data. Various studies have comparatively analyzed the prediction ability of these models with ML algorithms, such as Artificial Neutral Network (ANN), and have revealed that ML techniques forecast better than such models (Falat & Pancikova, 2015; Stamenković, Antanasijević, Ristić, Perić-Grujić, & Pocajt, 2015; Valipour, Banihabib, & Behbahani, 2013).

Recently, the advancement in ML techniques has played a critical role in decision making. For instance, ML algorithms such as decision tree, random forest, extreme gradient boosting, and SVR have played a critical role in modeling and forecasting. Although there have been studies evaluating the forecasting ability of classical statistical models and ML techniques, research comparing the forecasting ability of different ML algorithms by focusing on carbon emissions and more specifically on Australian's carbon emissions remains scarce. This warrants further empirical study. Therefore, in this chapter, we aim to evaluate the forecasting ability of the above four ML techniques by focusing on Australia's carbon emissions.

This study also contributes to the literature by employing macroeconomic variables, such as population size, economic growth, energy consumption, trade, financial development, foreign direct investment, and urbanization, which are mostly used as inputs to model carbon emissions (see Acheampong, 2018, 2019; Acheampong, Adams, & Boateng, 2019; Acheampong, Mary &, Boateng, 2020; Adams & Acheampong, 2019). To achieve robust and efficient estimates, this study uses high-frequency data. Finally, the outcome of this study will inform environmental policy-makers as to the best ML technique for forecasting carbon emissions in Australia. The rest of the chapter is organized as follows. In Section 9.2, the methodology and data are presented, while the results are discussed in Section 9.3. Conclusions and policy implications are presented in Section 9.4.

9.2 DATA AND METHODOLOGY

9.2.1 DATA

To compare the performance accuracy of the decision tree, random forest, extreme gradient boosting, and SVR, this study utilized a dataset for the period 1960–2018. Following Acheampong & Boateng (2019), we converted the data from an annual dataset to a quarterly dataset using the quadratic sum approach. Therefore, quarterly data, which range between 1960Q1 and 2018Q4, was used for the analysis. The output and input variables used are presented in Table 9.1. Economic growth, energy consumption, financial development, foreign direct investment, physical investment, population size, trade, and urbanization are used as the input variables for modeling carbon emissions.[2] All the variables were obtained from World Bank (2019).[3]

9.2.2 METHODOLOGY

9.2.2.1 Decision Trees

Decision trees (DTs) are non-parametric supervised learning techniques applied to regression and classification problems (Das, Naik, & Behera, 2020a; Das et al. 2020). These tree-based models are popular and advantageous in handling smaller datasets than neural network models. Through a repetitive process of splitting, the regression trees can yield a set of rules which can be used for prediction (Das et al., 2020b; Tso & Yau, 2007). The splitting process divides the sample into two or more homogeneous sets derived from the most important differentiator among the input variables. In the case of classification problems, metrics such as cross-entropy or the Gini index are used to decide strategic splits for DTs (Das et al., 2020; Xu et al., 2005). For regression problems, DTs normally use the mean squared error (MSE) criterion for splitting a node into two or more sub-nodes. That is, on each subset of data, the algorithm computes an MSE value and the tree with the least MSE is selected as a point of split. The concluding outcome comprises decision nodes and leaf nodes (Das, Naik, & Behera, 2020a, 2020b). By contrast with black-box models such as deep learning models, DTs are easy to understand and interpret because their rules can be visualized. These algorithms have been applied and attained success in many fields due to their efficiency and interpretability (Tsai & Chiou, 2009; Wu, 2009).

TABLE 9.1
Variables Used in This Study

Variables	Proxies	Mean	sd	min	max
Economic growth	GDP per capita (constant 2010 US$)	36,690.75	11,635.62	19,245.41	56,919.38
Energy use	Energy use (kg of oil equivalent per capita)	4805.486	808.1064	3063.554	5964.666
Population size	Population, total	1.70E+07	4115748	1.03E+07	2.50E+07
Financial development	Domestic credit to private sector (% of GDP)	64.06122	41.96032	17.65457	142.2841
Urbanization	Urban population growth (annual %)	1.644195	0.569214	0.768615	3.571777
Trade openness	Trade (% of GDP)	34.12007	6.410204	24.79318	45.7979
Physical capital investment	Gross capital formation (% of GDP)	27.83713	2.839597	22.39053	33.68189
Foreign direct investment	Foreign direct investment, net inflows (% of GDP)	2.171749	1.544092	-3.61882	7.005444
Carbon emissions	CO_2 emissions (kt)	252,419.8	92,195.4	88,202.35	394,792.9

9.2.2.2 Random Forests

A random forest (RF) is an ML algorithm that combines several DT models to effectively classify or predict an outcome (Breiman, 2001; Das, Naik, & Behera, 2020a, 2020c). This combination process, also termed "bootstrap aggregation" or "bagging," involves training each DT with a distinct set of observations through sample replacement. Samples which do not end up in a subset of training data during bagging are included with other subsets called "out-of-bag" (Rodriguez-Galiano et al., 2015). Bagging minimizes the variance of the base learner; however, it has minimal influence on the bias (Rodriguez-Galiano et al., 2015). Basically, on each node, there is a random selection of variables out of all possible variables, then the best split among the selected variables is determined, based on the lowest MSE. The final prediction is derived using an ensembling technique, which averages the predictions of the previous regression trees (Das et al., 2020). Due to the averaging of several trees, there is a considerably lower risk of overfitting (Breiman, 2001). The random sampling of training observations and random subsets of candidate variables for splitting nodes is a clear distinguishing factor of RF from DT.

9.2.2.3 Extreme Gradient Boosting

Extreme gradient boosting (XGBoost) is a scalable end-to-end tree-boosting algorithm (Chen & Guestrin, 2016). It can be applied to regression, ranking, and classification problems. This algorithm performs parallel tree learning using a novel sparsity-aware system (Chen & Guestrin, 2016). XGBoost employs the second-order Taylor expansion to approximate the loss function. This model has been known to outperform other ML models, and its success can be witnessed in numerous ML and data-driven competitions such as Kaggle (Dey et al., 2019; Rout et al., 2020). The fundamental factor behind the triumphs of XGBoost is its scalability in all circumstances (Chen & Guestrin, 2016). That is, using an optimal amount of resources, the algorithm yields state-of-the-art results when solving a wide range of problems. The implementation of this model is, thus, influenced by its high speed and performance.

9.2.2.4 Support Vector Regression

A support vector regression (SVR) is a type of support vector machine used for regression purposes and hence handles continuous values (Das et al., 2020; Dey et al., 2019). It follows the same principles as support vector classification, though with minimal modifications. The most important distinguishing factor between a simple linear regression model and SVR is that the former attempts to reduce the error rate while SVR tries to fit the error within a designated threshold. This kernel-based model has advantages in high dimensional spaces since its optimization does not rely on the dimensionality of the input area (Das et al., 2018; Drucker, Burges, Kaufman, Smola, & Vapnik, 1997). Moreover, SVR is a non-parametric tool and does not rely on distributions of the primary input and output variables. Although less popular than support vector classification (SVC), SVR has proven to be an effective technique in solving real-world scale problems (Awad & Khanna, 2015).

9.2.3 DATA DIVISION AND THE EXPERIMENTAL ENVIRONMENT

We compared the accuracies and computational costs of these four ML algorithms. Data used for training and validating the models were standardized to eliminate instances of one variable dominating another (Boateng, Pillay, & Davis, 2019), since the variables used in this study have different units (Acheampong & Boateng, 2019; Bannor & Acheampong, 2019). Eighty percent (188 quarters) of the data were used to train each model, while the remaining 20% (48 quarters) were used to validate the models. Similar data proportions were used by Morano et al. (2015); Lam et al. (2008); Acheampong & Boateng (2019); and Bannor & Acheampong (2019). For the hardware and software environment, we used an Intel i5-2520M (4) at 3.2 GHz CPU, and an 8 GB memory operating on Ubuntu 18.04.2 LTS. Two graphic processing units (GPUs) were used, an Intel 2nd Generation Core Proce, and an NVIDIA GeForce GTX 1050Ti with 8 GB memory. We used Spyder (Python 3.6.7) to write and execute the programming codes.

9.2.4 OPTIMIZATION OF HYPERPARAMETERS

As the goal of this study is to evaluate predictive ML models, there is the need to develop optimal models suitable for comparison purposes. We performed a grid search with ten-fold cross-validation on the hyperparameters of each model. This technique shuffles and resamples the training data into ten equal folds, fits the model on a combination of one set of hyperparameters on nine folds, and tests the model on the remaining fold (Bannor & Acheampong, 2019). A carefully tuned model is at lower risk of underfitting and overfitting problems. The best score function returns a combination of hyperparameters and associated arguments suitable for developing the model. For all models, we assessed their mean cross-validation scores (MCVs). The highest MCV is used as the basis to select the ideal combination of hyperparameters.

9.2.4.1 Parameter Tuning for the DT, RF, and XGBoost Algorithms

In prior experimentation, certain hyperparameters were deemed to influence the performance of the models significantly and hence were used in tuning the ML algorithms; in the case of the DT algorithm, five different maximum leaf nodes (none, 2, 3, 5, and 7), five maximum tree depths (1, 3, 5, 7, and 9), and five minimum samples for a leaf node (1, 3, 5, 7, and 9) over the ten folds of data results in 1250 models. We used a random state of zero for the DT regressor. For the RF algorithm, five numbers of estimators/trees (100, 150, 200, 250, and 300), five maximum tree depths (1, 3, 5, 7, and 9), and five minimum samples for a leaf node (1, 3, 5, 7, and 9) were specified. This also led to 1250 models. The hyperparameter arguments of the XGBoost, such as the number of estimators (100, 150, 200, 250, and 300), the number of maximum tree depths (1, 3, 5, 7, and 9), and the learning rates (0.01, 0.05, 0.1, 0.2, and 0.3), were also tuned, totaling 1250 models.

9.2.4.2 Parameter Tuning for the SVR Algorithm

For the SVR algorithm, six penalty "C" parameter arguments (10.0, 50.0, 100.0, 1000.0, 1050.0, and 1100.0), seven different gamma values (0.0005, 0.0001, 0.001,

0.01, 0.1, 1.0, and 10.0), and three types of kernels (radial basis function, linear function, and polynomial function) were experimented with over ten folds of the training dataset using the grid-search framework. In all, 1260 models were developed for the SVR. After the grid search with the ten-fold cross-validation exercise, we selected the ideal hyperparameter arguments for each algorithm based on their MCV scores.

9.2.5 Performance Metrics

We assessed the accuracies of each algorithm in predicting the 48 carbon emission data points in the test dataset. By evaluating the deviations between the predicted and actual emissions, models with lower errors were ranked high in terms of accuracy. The root mean squared error (RMSE), the coefficient of determination (R^2), and the mean absolute percentage error (MAPE) were used in comparing the levels of deviation among the four ML algorithms. We also assessed the computational efficiency of the four models. In particular, the elapsed time taken during the grid search with a ten-fold cross-validation process on each algorithm was measured.

9.3 RESULTS AND DISCUSSION

9.3.1 Development and Validation of the DT Model

After performing the grid-search with ten-fold cross-validation, the DT model with the parameter combination of a maximum tree depth of 9, maximum leaf nodes as "none," and minimum samples for a leaf node as 1 had the highest MCV score of 0.975899. These parameters and their arguments were used in configuring the DT model, before training and validating it. This relationship between the hyperparameters and the MCV scores during the cross-validation process is depicted in Figure 9.1. It can be observed that the higher the depth of the tree, the higher the maximum leaf nodes grow, and hence a more accurate DT model is obtained. That is, the "none" argument of the maximum leaf nodes parameter suggests an unlimited number of nodes relative to the decrease in errors. On the other hand, a higher tree depth corresponds well with a few minimum samples required for a leaf node.

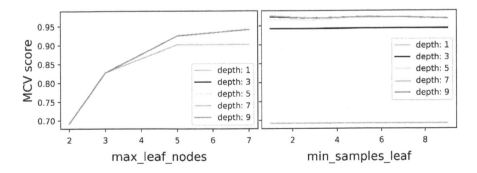

FIGURE 9.1 Optimization of hyperparameters for DT models.

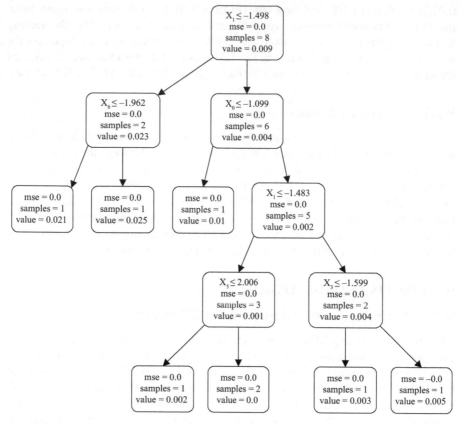

FIGURE 9.2A Section of DT model plot.

Sections of the validated DT model are shown in Figures 9.2a and 9.2b. Negligible MSEs can be observed at each internal and leaf node, and hence it is suitable to compete with the other three ML models. The variables X_0 to X_7 within the internal nodes tally with the features in the input dataset; thus, the first feature "credit to private sector" is X_0, "GDP" is X_1, "physical investment" is X_2, "population" is X_3, "trade" is X_4, "urbanization" is X_5, "energy use" is X_6, and "FDI" is X_7.

9.3.2 Development and Validation of the RF Model

The RF model with the parameter combination of a maximum tree depth of 9, with minimum samples for a leaf node as 1, and with the number of estimators/trees at 250 had the highest MCV score of 0.984858, attained after evaluating all 1250 RF models from the grid search with a ten-fold cross-validation process. Corroborating the findings from the DT model, higher tree depths are inversely related to the minimum samples required for a leaf node. For instance, in Figure 9.3, the maximum number 9 of tree depth corresponds to a minimum sample of 1 required for a leaf node.

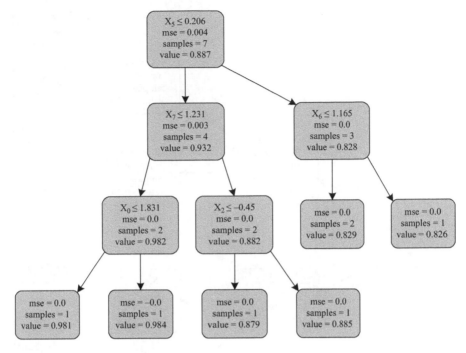

FIGURE 9.2B Section of DT model plot.

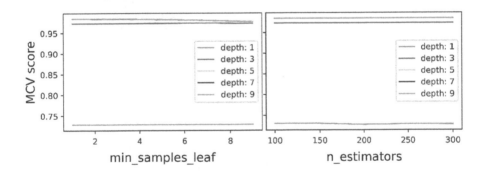

FIGURE 9.3 Optimization of hyperparameters for RF models.

Figure 9.4 shows the DT model configured with the best hyperparameter argu-ments for the grid search, which is then trained with 188 quarters of emissions data and validated on 48 quarters. Similar to the DT model, the RF model's (Figure 9.4) variables of X_0 to X_7 in the root and internal nodes corresponds to the features "credit to the private sector" to "FDI" respectively. In Figure 9.4, these features are consid-ered in constructing questions while accessing a random set of training observations. The number of samples on each internal node are also depicted, and associated errors on the predictions or decisions on each node are evaluated with the MSE criterion. Our RF model consists of 250 DTs combined into a single predictive model. We plot

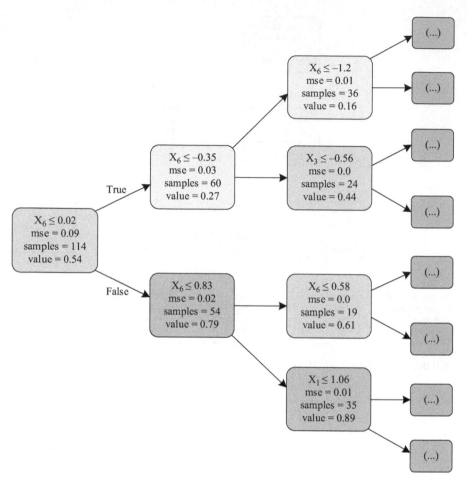

FIGURE 9.4   RF model plot of 250th tree at a maximum depth of 2.

only the 250th DT at a maximum depth of 2 due to brevity. Using the wisdom of the crowd, the resulting outcomes or predictions of the RF model are made in consultation with other DTs by averaging out their predictions.

In Figure 9.4, the first node is the root node, while the remaining nodes are the internal nodes. As this tree is only shown at a depth of 2, the leaf nodes are hard to visualize (due to brevity) as the complete maximum depth of a tree is 9 in this study. For the 250th tree at a maximum depth of 2, only three features were important according to the model and hence are used to make decisions. These three were GDP (X_1), population (X_3), and energy use (X_6).

9.3.3 Development and Validation of the XGBoost Model

Next, MCV scores of 1250 XGBoost models were evaluated after the grid search with ten-fold cross-validation. The model with the hyperparameter combination of a

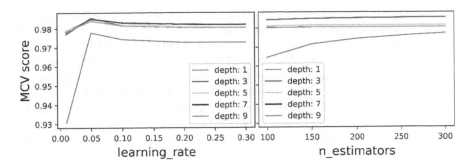

FIGURE 9.5   Optimization of hyperparameters for XGBoost models.

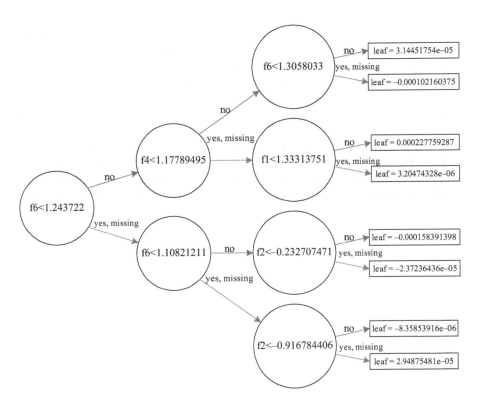

FIGURE 9.6   XGBoost model plot at 300th tree.

0.05 learning rate, a maximum tree depth of 3, and the number of estimators/trees at 300 attained the highest MCV score of 0.985607. Figure 9.5 shows that, irrespective of the maximum number of the tree depth, higher MCV scores were associated with the 0.05 learning rate. This scenario is evident, as an upwards kink can be observed at the 0.05 learning rate. These ideal parameters, as determined by the grid-search framework, were used to build, calibrate, and test the model (Figure 9.6).

As shown in Figure 9.6, we plot the 300th tree of the XGBoost model. While "X_n" is used as a feature in the DF and RF models, "fn" is used for XGBoost. Specifically, X_0 to X_7 in the DT and RF models correspond to f0 to f7 in the XGBoost model. At the 300th tree, four features (GDP, physical investment, trade, and energy use) were deemed important by the model and hence used in making the predictions. The important features used to make the decisions often vary from tree to tree and, in other instances, some trees may have the same features. It can also be observed that the influence of the maximum tree depth of 3 matches the three layers of internal nodes in Figure 9.6. The questions or conditions on the internal nodes suggest which data could be sieved through. The scores on the leaf nodes are also termed "weights."

9.3.4 DEVELOPMENT AND VALIDATION OF THE SVR MODEL

After performing the grid search with ten-fold cross-validation, 1260 SVR models were developed. The model with hyperparameters of a C of 50.0, a gamma of 0.001, and an RBF kernel reached the highest MCV score of 0.967804. Considering the difficulties of illustrating more than three features in a 3D chart, we show separately for each kernel the MCV scores attained in association with other hyperparameters (Figure 9.7). During the grid-search process, it can be observed that an increase in the gamma does not have a significant influence on the linear and polynomial SVR models.

On the other hand, a significant inverse relationship is seen between the gamma and the MCV score of the RBF models. This is because the gamma regulates the influence of new features on the decision boundary. Hence, a too large gamma could result in overfitting, and no amount of C could avoid it. In order for the models to

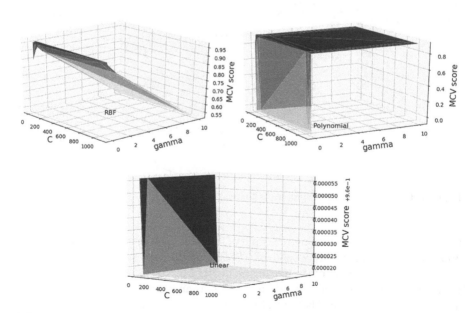

FIGURE 9.7   3D grid-search results for SVR models.

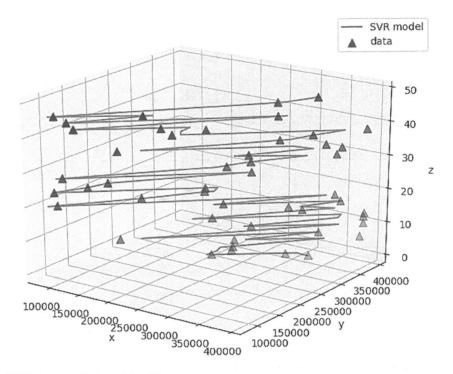

FIGURE 9.8   SVR model in 3D view.

predict accurately, the gamma is lowered to an optimal level to capture the complexity of the data. Figure 9.8 shows the trained and validated RBF-SVR model which
was configured with the ideal parameters derived from the grid-search process. The
SVR model shows promising results, as the best line of fit accurately passes through/
near the test data points.

9.3.5 Performance Evaluation of Model

We compared the amount of observed variance that each validated model explained
(Figure 9.9). The DT model explained almost 100% of the observed variance, putting
its R^2 at 99.71%, and hence having its data points almost perfectly fitted on the
regression line. The RF model followed closely with 99.14%. The XGBoost model
also accounted for 98.88% of the observed variance around its mean. Though the
SVR model had the lowest R^2 in this study, its ability to explain 97.42% of the
observed variance leaves less than 3% of the variance accounted for by the model.

   In terms of accuracy, the tree-based models had the lowest MAPE and RMSE, as
shown in Table 9.2. In all, the DT model produced the most accurate predictions. On
the other hand, the kernel-based model, RBF-SVR, had comparatively higher errors.
This could be observed in Figure 9.8 as the SVR model only passed through/near a
few of the test data points. The DT and SVR models were more computationally
efficient than the other two models. Overall, the DT model ranked top in all spheres
of performance assessment in this study.

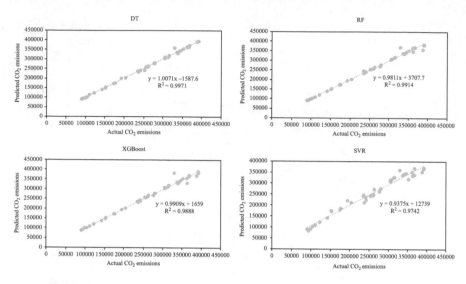

FIGURE 9.9 Scatter diagram of actual and predicted CO_2 emissions.

TABLE 9.2
Performance Evaluation of ML Models

Model	MAPE (%)	RMSE[†]	R^2	Elapsed Time	Ranking Accuracy	Time
DT	1.1076	5145.3328	0.9971	2.2 s for 1250 fits	1st	1st
RF	1.4812	8834.8885	0.9914	1.9 min for 1250 fits	2nd	4th
XGBoost	1.6228	9948.4937	0.9888	39.2 s for 1250 fits	3rd	3rd
SVR	4.8045	15,823.5338	0.9742	2.6 s for 1260 fits	4th	2nd

[†] CO_2 emissions (metric tons per capita).

9.4 CONCLUDING REMARKS

The development of a robust, high-quality, and accurate model for forecasting carbon emissions is a prerequisite for providing insights into environmental policies for achieving the Paris Agreement on climate change. Research comparing the forecasting ability of different ML algorithms, by focusing on carbon emissions and more specifically on Australian's carbon emissions, remains scarce. This warrants further research. This chapter has applied different ML techniques, such as DT, RF, XGBoost, and SVR, to model Australia's carbon emissions. The findings indicate that the DT has a high coefficient of determination (R^2) of 99.71%, followed by the RF with an R^2 of 99.14%, XGBoost with an R^2 of 98.88%, and SVR with an R^2 of 97.42%. In terms of accuracy, the tree-based models had the lowest MAPE and RMSE. Overall, the DT model produced the most accurate predictions. On the other hand, the kernel-based model, RBF-SVR, had comparatively higher errors. For computational efficiency, the DT and SVR models were more efficient than XGBoost and RF. Comparatively, the DT model ranked first among the other ML techniques utilized in this study, based on the performance assessment metrics. The implication is that for accurate prediction and effective planning of policies to mitigate climate change, Australia's environmental planners should incorporate a decision tree as one of the techniques for modeling and forecasting carbon emissions. Some ML algorithms could have performed better in terms of their computational efficiency; however, the combination and choice of certain parameters for such algorithms during the grid-search process can be computationally laborious in general and hence, to a marginal extent, could limit our results in those aspects. Other robust and novel ML algorithms could be employed to explain the 100% variance in the dataset. As this study was for regression, future research could investigate the classification problems pertaining to Australia's carbon emissions.

REFERENCES

Acheampong, A. O. (2018). Economic growth, CO_2 emissions and energy consumption: What causes what and where? *Energy Economics*, 74, 677–692. doi:https://doi.org/10.1016/j.eneco.2018.07.022

Acheampong, A. O. (2019). Modelling for insight: Does financial development improve environmental quality? *Energy Economics*, 83, 156–179. doi:https://doi.org/10.1016/j.eneco.2019.06.025

Acheampong, A. O., & Boateng, E. B. (2019). Modelling carbon emission intensity: Application of artificial neural network. *Journal of Cleaner Production*, 225, 833–856. doi:https://doi.org/10.1016/j.jclepro.2019.03.352

Acheampong, A. O., Adams, S., & Boateng, E. (2019). Do globalization and renewable energy contribute to carbon emissions mitigation in Sub-Saharan Africa? *Science of the Total Environment*, 677, 436–446. doi:https://doi.org/10.1016/j.scitotenv.2019.04.353

Acheampong, A. O., Amponsah, M., & Boateng, E. (2020). Does financial development mitigate carbon emissions? Evidence from heterogeneous financial economies. *Energy Economics*, 88, 104768. doi:https://doi.org/10.1016/j.eneco.2020.104768

Adams, S., & Acheampong, A. O. (2019). Reducing carbon emissions: The role of renewable energy and democracy. *Journal of Cleaner Production*, 240, 118245. doi:https://doi.org/10.1016/j.jclepro.2019.118245

Auffhammer, M. (2007). The rationality of EIA forecasts under symmetric and asymmetric loss. *Resource and Energy Economics*, 29(2), 102–121.

Awad, M., & Khanna, R. (2015). Support vector regression. In: M. Awad & R. Khanna (Eds.), *Efficient learning machines: Theories, concepts, and applications for engineers and system designers* (pp. 67–80). Berkeley, CA: Apress.

Bannor, B. E., & Acheampong, A. O. (2019). Deploying artificial neural networks for modeling energy demand: international evidence. *International Journal of Energy Sector Management*, 14(2), 285–315. doi:10.1108/IJESM-06-2019-0008

Boateng, E. B., Pillay, M., & Davis, P. (2019). Predicting the level of safety performance using an artificial neural network. In T. Ahram, W. Karwowski, & R. Taiar (Eds.), *Human systems engineering and design* (pp. 705–710). Cham: Springer International Publishing.

Breiman, L. (2001). Random forests. *Machine Learning*, 45(1), 5–32.

Chen, T., & Guestrin, C. (2016). XGBoost: A scalable tree boosting system. In *Proceedings of the 22nd ACMSIGKDD International Conference on Knowledge Discovery and Data* (pp. 785–794).

Climate Council. (2019). *Compound cost: How climate change is damaging Australia's economy.* Retrieved from www.climatecouncil.org.au/resources/compound-costs-how-climate-change-damages-australias-economy/.

Das, H., Naik, B., & Behera, H. S. (2018). Classification of diabetes mellitus disease (DMD): A data mining (DM) approach. In: P. Pattnaik, S. Rautaray, H. Das, & J. Nayak (Eds.), *Progress in computing, analytics and networking* (pp. 539–549). Singapore: Springer.

Das, H., Naik, B., & Behera, H. S. (2020a). An experimental analysis of machine learning classification algorithms on biomedical data. In: S. Kundu, U. Acharya, & S. Mukherjee (Eds.), *Proceedings of the 2nd international conference on communication, devices and computing* (pp. 525–539). Singapore: Springer.

Das, H., Naik, B., & Behera, H. S. (2020b). A hybrid neuro-fuzzy and feature reduction model for classification. *Advances in Fuzzy Systems*, 2020(1), 1–15.

Das, H., Naik, B., & Behera, H. S. (2020c). Medical disease analysis using neuro-fuzzy with feature extraction model for classification. *Informatics in Medicine Unlocked*, *18*, 100288.

Das, H., Naik, B., Behera, H. S., Jaiswal, S., Mahato, P., & Rout, M. (2020). Biomedical data analysis using neuro-fuzzy model with post-feature reduction. *Journal of King Saud University-Computer and Information Sciences.*

Dey, N., Das, H., Naik, B., & Behera, H. S. (Eds.). (2019). *Big data analytics for intelligent healthcare management.* London: Academic Press.

Drucker, H., Burges, C. J., Kaufman, L., Smola, A. J., & Vapnik, V. (1997). *Support vector regression machines.* Paper presented at *the Advances in Neural Information Processing Systems*, Denver, CO.

Falat, L., & Pancikova, L. (2015). Quantitative modelling in economics with advanced artificial neural networks. *Procedia Economics and Finance*, 34, 194–201. doi:https://doi.org/10.1016/S2212-5671(15)01619-6

Gallo, C., Contò, F., & Fiore, M. (2014). A neural network model for forecasting CO_2 emission. *Agris on-line Papers in Economics and Informatics*, 6(2), 31.

Lam, K. C., Yu, C. Y., & Lam, K. Y. (2008). An artificial neural network and entropy model for residential property price forecasting in Hong Kong. *Journal of Property Research*, 25(4), 321–342.

Morano, P., Tajani, F., & Torre, C. M. (2015). Artificial intelligence in property valuations an application of artificial neural networks to housing appraisal. *Advances in Environmental Science and Energy, Planning*, 23–29.

Ndevr Environmental. (2019). *Tracking 2019 Q3 report.* Retrieved from https://ndevrenvironmental.com.au/tracking-2-degrees-fy2019-q3/

O'Neill, B. C., & Desai, M. (2005). Accuracy of past projections of US energy consumption. *Energy Policy*, 33(8), 979–993.

Rodriguez-Galiano, V., Sanchez-Castillo, M., Chica-Olmo, M., & Chica-Rivas, M. (2015). Machine learning predictive models for mineral prospectivity: An evaluation of neural networks, random forest, regression trees and support vector machines. *Ore Geology Reviews*, 71, 804–818.

Rout, M., Jena, A. K., Rout, J. K., & Das, H. (2020). Teaching–learning optimization based cascaded low-complexity neural network model for exchange rates forecasting. In: S. Satapathy, V. Bhateja, J. Mohanty, & S. Udgata (Eds.), *Smart intelligent computing and applications* (pp. 635–645). Singapore: Springer.

Stamenković, L. J., Antanasijević, D. Z., Ristić, M. Đ., Perić-Grujić, A. A., & Pocajt, V. V. (2015). Modeling of methane emissions using artificial neural network approach. *Journal of the Serbian Chemical Society*, 80(3), 421–433.

Tsai, C.-F., & Chiou, Y.-J. (2009). Earnings management prediction: A pilot study of combining neural networks and decision trees. *Expert Systems with Applications*, 36(3), 7183–7191.

Tso, G. K., & Yau, K. K. (2007). Predicting electricity energy consumption: A comparison of regression analysis, decision tree and neural networks. *Energy*, 32(9), 1761–1768.

Valipour, M., Banihabib, M. E., & Behbahani, S. M. R. (2013). Comparison of the ARMA, ARIMA, and the autoregressive artificial neural network models in forecasting the monthly inflow of Dez dam reservoir. *Journal of Hydrology*, 476, 433–441. doi:https://doi.org/10.1016/j.jhydrol.2012.11.017

World Bank. (2019). World development indicators. Retrieved from http://databank.worldbank.org/data/reports.aspx?source=world-development-indicators#

Wu, D. (2009). Supplier selection: A hybrid model using DEA, decision tree and neural network. *Expert Systems with Applications*, 36(5), 9105–9112.

Xu, M., Watanachaturaporn, P., Varshney, P. K., & Arora, M. K. (2005). Decision tree regression for soft classification of remote sensing data. *Remote Sensing of Environment*, 97(3), 322–336.

Yin, M.-S., & Tang, H.-W. V. (2013). On the fit and forecasting performance of grey prediction models for China's labor formation. *Mathematical and Computer Modelling*, 57(3), 357–365. doi:https://doi.org/10.1016/j.mcm.2012.06.013

Zhao, X., & Du, D. (2015). Forecasting carbon dioxide emissions. *Journal of Environmental Management*, 160, 39–44. doi:https://doi.org/10.1016/j.jenvman.2015.06.002

NOTES

1 https://ndevrenvironmental.com.au/tracking-2-degrees-fy2019-q3/.
2 See Acheampong and Boateng (2019).
3 https://data.worldbank.org/.

10 Prediction Analysis of Exchange Rate Forecasting Using Deep Learning-Based Neural Network Models

Dhiraj Bhattarai, Ajay Kumar Jena, and Minakhi Rout

CONTENTS

10.1 INTRODUCTION

Currency exchange rate forecasting is a crucial technique to gauge the foreign exchange rate of various currencies using different algorithms and models. The accurate prediction of forex values can be fruitful to fiscal institutions dealing with import and export business globally. Hence, economists, investors, and researchers are always keen to anticipate the future forex value to continue to rely upon the business. Because of the dynamic nature of the exchange rate, estimating succeeding values with 100% accuracy is impossible, so several linear statistical methods have been developed and implemented over the decades. But, the introduction of artificial neural networks (ANNs) has improved performance by adjusting the model at every step with calculated errors, although the use of the modified single layered, multi-layered, or complex architecture of neural networks has reduced the occurrence of error and risk over high-frequency financial time-series data. The sophisticated ANN model can handle time-series data and can forecast long-term exchange rate prediction with more accuracy.

In the field of financial forecasting, much research has been done and implemented to step up the accuracy. The standard deviation of forex in the logarithm rate over a linear model [1], daily minimum and maximum weighted average price are experimentally better than the daily close price forecast. Estimation of forex [2] values using three ANN techniques (i.e., Standard Back-Propagation (SBP), Back-Propagation with Bayesian Regularization (BPR), and Scaled Conjugate Gradient (SCG)) performed exceptionally better than the traditional ARIMA model. Also, [3] a single layer FLANN and Lagueree Polynomial Equation (LAPE) performed better than ARIMA. Technical analysis applying a neural network [4] on the USD/GBP predicted with a low mean squared error, and in conjunction with other financial and technical forms. Generally short-term forex predictions are more feasible [5] than long-term predictions. The accuracy and efficiency of any model can be improved using weekly, monthly, and quarterly data samples. In comparison with time-series input values [6], one-week-ahead prediction outperforms one-day-and-a-month-ahead prediction because of the appropriate fluctuation of the exchange rate in weekly data. Hybrid models have been implemented to improve overall performance [7, 8], aggregating On-Line Sequential Extreme Learning Machine (OL-SELM) with Krill Herd (KH) optimizing feature reduction [9]. The accuracy of the ensemble model increases with the increased performance of individual measures. By applying the Gaussian Mixture Model Initialized Neuro-Fuzzy (GMMINF) technique [10] for data partitioning on historical observations, performance can be enhanced. A hybrid ANN, implementing a genetically optimized adaptive model [11] and a local regression model [12], predicts accurately by combining a genetic algorithm and a training method based on an Extended Kalman Filter (EKF) to develop the structure and train a multiple layer network. Another hybrid network [13] having two parallel systems (Functional Link Artificial Neural Network (FLANN) and Least Mean Square (LMS)) outperforms individual FLANN and LMS models. Recurrent Cartesian Genetic Programming [14] can decide whether to use feed-forward or recurrent connection; accuracy is increased with feedback path numbers and improves network capability. Similarly, a Convolutional-Recurrent Neural Network (RNN) [15] based on Deep RNN and Deep Convolutional Neural Network improves the accuracy of a deep learning algorithm for time-series data. The recently introduced Deep Belief approach [16, 17] has some improvements and results that are better than a typical Feed-Forward Neural Network (FFNN), which can model time series using Continuous Restricted Boltzman Machines. Also in the stock market, 33 technical indicators (based on daily stock prices, such as open, high, low, and close prices) can be extracted to address technical indicator feature selection and identification of the relevant technical indicator by using a Bourta feature selection technique [18]. Adaptive Smoothing Network [19] adjusts learning parameters automatically and tracks signals under dynamic varying environments to improve convergence.

Similarly, a regularized dynamic Multi-Layer Perceptron (MLP) [20, 21] applies an immune algorithm and an old statistical technique, Simple Moving Average (SMA) and Auto-Regressive Integrated Moving Average (ARIMA). The Support Vector Machine (SVM) [22] demonstrates more accurate prediction than ARIMA. One-month-ahead forecasting [23] using a single layer ANN model having non-linear input trained with historical data yields excellent results. Bootstrap [24], ensemble multiple learning models, and a combination of each output using a combination function, outperforms a traditional single model (e.g., NN and SVM). A two-stage hybrid model [25] incorporates parametric (ARIMA and Vector Auto-Regressive) and non-parametric (MLP and SVR) techniques to identify dependent and independent variable values, and estimates the mathematical model's parameters. SVR performs better than MLP in solving the quadratic programming problem and has a global optimum solution. Previously, most work used a simple ANN architecture with a single or multiple or hybrid neural network using various algorithms. However, these networks struggle with the long-term dependency problem. Thus, ANN, having memory blocks or states in which previous information persists, can handle the long-term dependency problem. To overcome these problems a novel architecture can be used, combining a Generative Adversarial Network (GAN) [26] and MLP as a discriminator, and Long-Short Term Memory (LSTM) as a generator for estimating the closing price of the stock market. Generators mine the data distribution and discriminators discriminate real stock data and generated data using LSTM and MLP, respectively. Machine learning approaches [27], like FFNN, simple RNN, Gated Recurrent Unit (GRU), and LSTM, have been used to predict Nepalese rupees against various currencies. LSTM estimated accurately over three other models for different input parameters (open, close, high, and low). The ability of the LSTM network [28] to store historical data to forecast future values minimizes error in the model. In this chapter, the exchange rate estimation of two ANN methods, LSTM and GRU, is performed. The implementation and working of this special RNN model are explained.

Section 10.2 describes the methodology and includes the working of the model, performance measuring parameters, and data preparation methods. Section 10.3 presents the results with simulation, and Section 10.4 concludes.

10.2 METHODOLOGY

Humans understand things based on their previous knowledge. We do not reset our brains and start thinking again from the beginning; our thoughts have persistence. An RNN is a mysterious kind of artificial neural network having a loop in it to allow information to persist. They are the same as normal neural networks, but have multiple copies of similar networks. In this chapter, the GRU network, a special type of recurrent network introduced by Kyunghyun Cho et al. in 2014 [29], is implemented to predict the foreign currency exchange rates. A GRU model works similarly to LSTM [30] but has fewer gates and modified networks. It works exceptionally well on various problems and is now widely implemented. These networks are designed to discard long-term dependency problems. Remembering previous information for a long time is their default behavior, not something they struggle to retain. An output gate is missing in this approach and the modified network consists of two gates, known as an update and a reset gate, that make it simpler in architecture and faster in execution; it also reduces calculation complexity. The remaining gates handle the function of the missing gate. An update gate acts similarly to a forget and input gate: it decides which information to add and which to throw away. How many previous values to discard is decided by the reset gate. In this way, it can keep only the relevant data for a prediction problem and can forget non-relevant information. A GRU network has memory blocks in the place of neurons. These blocks contain gates that decide which hidden state a value must choose. The recurrent networks take inputs sequentially and transformed them into machine-readable vectors, while the previous output of the hidden state acts as a memory. Input and the previous hidden state are fed into the network and activated using sigmoid and tanh activation functions, which give outputs between 0 and 1 from sigmoid, and -1 and 1 from tanh. The mathematics behind this model is given below.

The input variables from the daily exchange rate dataset are passed through the GRU block (Figure 10.1). The update gate z_t takes input data x_t and hidden state data h_{t-1} as input, which are then multiplied with their respective weight matrix (W_z and U_z) and the obtained result is activated using the sigmoid activation function (σ):

$$\left(\sigma = \frac{1}{\left(1+e^{-x}\right)} \right)$$

$$z_t = \sigma\left(W_z \cdot x_t + U_z \cdot h_{t-1} + b_z\right) \qquad (10.1)$$

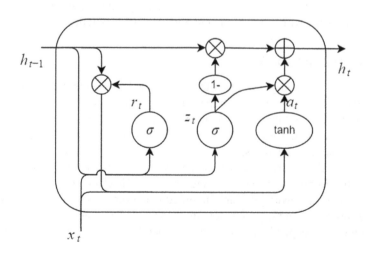

FIGURE 10.1   GRU model.

Again, the reset gate r_t takes inputs from the input dataset and hidden state data, which are then multiplied by weight matrices W_r and U_r respectively and then activated using the sigmoid function:

$$r_t = \sigma\left(W_r \cdot x_t + U_r \cdot h_{t-1} + b_r\right) \tag{10.2}$$

Now, to change the current memory content, activated aggregate calculation a_t is carried out using previous information and the current input data. In this step, the input data x_t and the product of the hidden state data and the value of the reset gate are multiplied with the corresponding weight variables W_a and U_a. The obtained result is then activated using the hyperbolic tangent (tanh) activation function:

$$\left(\tanh = \frac{\left(e^x - e^{-x}\right)}{\left(e^x + e^{-x}\right)} \right)$$

$$a_t = \tanh\left(W_a \cdot x_t + U_a \cdot \left(r_t \cdot h_{t-1}\right) + b_a\right) \tag{10.3}$$

Finally, the memory or the current hidden state data is changed with the help of the above calculated results:

$$h_t = \left(1 - z_t\right) \cdot h_{t-1} + z_t \cdot a_t \tag{10.4}$$

The calculated h_t of this state works as the memory or previous state data (h_{t-1}) for the next state. As we know, the sigmoid function has values between 0 and 1; the discard and update of the data depends upon the values of z_t and $(1 - z_t)$. These two variables are interlinked; the greater the value of one variable, the less will be the value of another, and vice versa.

10.2.1 PERFORMANCE MEASURE

We employ the Mean Absolute Percentage Error (MAPE), the Root Mean Square Error (RMSE), the Coefficient of Determination (R^2), Average Relative Variance (ARV), Theil's U, and Directional Accuracy (DA) to evaluate the performance characteristics of the proposed neural network models as follows:

$$\text{MAPE} = \left(\frac{1}{n}\sum \frac{|A - P|}{|A|}\right) * 100 \tag{10.5}$$

$$\text{RMSE} = \sqrt{\left(\frac{1}{n}\sum\left(A - P\right)^2\right)} \tag{10.6}$$

where A is the actual value, P is the predicted value, and n is the number of the predictions carried out. The values of MAPE and RMSE are greater than zero, and the closer the result is to zero, the more accurate will be the performance.

Another equation we apply for the performance measure is the coefficient of determination, denoted by R^2 and read as "R-squared." This is the proportion calculated from actual and predicted variables:

$$\bar{A} = \frac{1}{n}\sum_{i=1}^{n} A_i$$

$$SS_{tot} = \sum_{i}\left(A_i - \bar{A}\right)^2$$

$$SS_{res} = \sum_{i}\left(A_i - P_i\right)^2 \qquad (10.7)$$

$$R^2 = 1 - \frac{SS_{res}}{SS_{tot}}$$

where A and P are actual and predicted variables, $\bar{A}$ is the mean of A, and ss_{tot} and ss_{res} are the total sum of squares and the residual sum of squares. The value of R^2 lies between 0 and 1, and the value closest to 1 is the more accurate.

$$DA = \frac{100}{n}\sum_{i=1}^{n} d_i \quad \text{where } d_i = \begin{cases} 1, & \text{if } \left(P_i - P_{i-1}\right)\left(A_i - A_{i-1}\right) \ge 0 \\ 0, & \text{otherwise} \end{cases} \qquad (10.8)$$

where d_i is the direction accuracy data, P_i and A_i are the predicted data and actual data. This measures the number of correct decisions made when forecasting whether the next values will increase or decrease during the subsequent steps. The values of DA fall between 0 and 100; the closer the values are to 100, the more accurate the prediction of the model is. This performance-measuring approach is more important to stock market analysis because of the direction of the time-series data.

$$\text{Theil's U} = \frac{\sqrt{\frac{1}{2}\sum_{i=1}^{n}\left(A_i - P_i\right)^2}}{\sqrt{\frac{1}{2}\sum_{i=1}^{n} A_i^2} + \sqrt{\frac{1}{2}\sum_{i=1}^{n} P_i^2}} \qquad (10.9)$$

The model is unable to predict if the value of Theil's U approaches zero. Another performance measure, the ARV, is defined as:

$$ARV = \frac{\sum_{i=1}^{n}\left(P_i - A_i\right)^2}{\sum_{i=1}^{n}\left(P_i - \bar{A}\right)^2} \qquad (10.10)$$

The model is practical if the ARV value is between 0 and 1, where the closer the value is to zero, the better is the result.

10.2.2 DATA PREPARATION

The historical foreign currency exchange rates data downloaded from The Financial Forecast Center (www.forecasts.org), for 1570 daily forex samples of five different currencies (the Australian dollar (AUD), the British pound (GBP), the Canadian dollar (CAD), the Indian rupee (INR), and the Japanese yen (JPY)) against 1 US dollar (USD), were obtained over the period January 2012 to April 2018. These daily dataset samples were normalized in the range of [0, 1], dividing each value by the maximum value. The input patterns and the corresponding output values are extracted by using the concept of sliding window size, one step at a time. N number of data samples and W sliding window size generate the total number of patterns equal to $(N - W + 1)$. In this feature extraction technique, the number of input variables (or the column size of the extracted values) is taken similarly to the size of the sliding windows, in which the first column of the extracted dataset starts from the first index of the values, the second column starts from the second index, and so on. The generated pattern to input the model is further divided into two different datasets; the first 80% of samples are used for training purposes and the remaining 20% for testing.

10.3 RESULTS AND SIMULATIONS

Tables 10.1–10.15 show the prediction performance of two different models: the LSTM model and the GRU model, over various days-ahead predictions. The historical forex data of five different currencies are taken and applied to the proposed model. The resulted outputs are then compared with the actual output to measure the performance characteristics of both approaches. Although monetary exchange values are dynamic in nature, sudden increments or decrements of an exchange rate is quite difficult to handle using traditional networks; they also require more time to understand the pattern and update the weights to adjust the model accordingly. As the recurrent neural networks can remember previous information, these two special kinds of RNN can handle complex data with long term dependency and execute them in fewer iterations. The results of the LSTM and GRU models are tabulated by taking different window sizes (i.e., 7, 10, and 13). While increasing the number of the days-ahead prediction, the values of the performance measuring parameters also increase.

10.3.1 FOR SLIDING WINDOW SIZE 7

According to Tables 10.1–10.5, the GRU model has performed better than the LSTM model. Some of the performance measuring parameters have similar values for both models but, as a whole, the GRU approach gives more accurate results for the given samples of the daily dataset for sliding window size 7.

TABLE 10.1
The One-Day-Ahead Prediction Performance of the LSTM and GRU Models

Currency	LSTM Model						GRU Model					
	MAPE	RMSE	R²	ARV	Theil's	DA	MAPE	RMSE	R²	ARV	Theil's	DA
AUD	0.4298	0.0041	0.9480	0.0550	0.0027	57	0.3724	0.0036	0.9585	0.0421	0.0024	57
CAD	0.3592	0.0061	0.9773	0.0231	0.0023	59	0.3397	0.0058	0.9792	0.0212	0.0022	58
GBP	0.4758	0.0079	0.9785	0.0219	0.0030	54	0.4068	0.0068	0.9840	0.0164	0.0026	54
INR	0.1889	0.1697	0.9762	0.0238	0.0013	50	0.1880	0.1694	0.9763	0.0240	0.0013	49
JPY	0.4238	0.6241	0.9339	0.0661	0.0028	50	0.4112	0.6097	0.9369	0.0626	0.0027	50

TABLE 10.2
The Three-Days-Ahead Prediction Performance of the LSTM and GRU Models

Currency	LSTM Model						GRU Model					
	MAPE	RMSE	R²	ARV	Theil's	DA	MAPE	RMSE	R²	ARV	Theil's	DA
AUD	0.7073	0.0068	0.8527	0.1505	0.0044	55	0.7094	0.0068	0.8525	0.1454	0.0044	55
CAD	0.6392	0.0104	0.9331	0.0728	0.0040	57	0.6299	0.0103	0.9347	0.0697	0.0697	57
GBP	0.7754	0.0130	0.9419	0.0607	0.0050	53	0.7272	0.0122	0.9490	0.0519	0.0046	54
INR	0.3417	0.2934	0.9272	0.0709	0.0023	49	0.4434	0.3595	0.8907	0.1032	0.0028	49
JPY	0.7551	1.0412	0.8134	0.1817	0.0047	50	0.7265	1.0070	0.8255	0.1741	0.0045	51

TABLE 10.3

The Five-Days-Ahead Prediction Performance of the LSTM and GRU Models

Currency	LSTM Model						GRU Model					
	MAPE	RMSE	R^2	ARV	Theil's	DA	MAPE	RMSE	R^2	ARV	Theil's	DA
AUD	0.9116	0.0089	0.7477	0.2584	0.0058	58	**0.8951**	**0.0087**	**0.7599**	0.2366	**0.0057**	**60**
CAD	**0.8557**	**0.0137**	**0.8843**	0.1340	**0.0053**	56	0.8952	0.0143	0.8746	0.1334	0.0055	56
GBP	0.9870	0.0165	0.9060	0.1014	0.0063	57	**0.9074**	**0.0149**	**0.9239**	**0.0796**	**0.0057**	**58**
INR	0.4801	0.3939	0.8688	0.1252	0.0030	54	**0.4202**	**0.3612**	**0.8897**	**0.1076**	**0.0028**	53
JPY	0.9572	1.3271	0.6969	**0.2882**	0.0060	**57**	0.9516	1.3142	0.7028	0.2883	**0.0059**	54

TABLE 10.4

The Seven-Days-Ahead Prediction Performance of the LSTM and GRU Models

Currency	LSTM Model						GRU Model					
	MAPE	RMSE	R^2	ARV	Theil's	DA	MAPE	RMSE	R^2	ARV	Theil's	DA
AUD	1.1776	**0.0114**	**0.5866**	**0.3865**	**0.0074**	56	1.1946	0.0116	0.5738	0.3981	0.0076	**57**
CAD	1.0249	**0.0164**	**0.8341**	0.1980	**0.0063**	54	1.0274	0.0164	0.8332	0.1837	0.0064	**55**
GBP	1.2420	0.0210	0.8486	0.1631	0.0080	52	**1.2090**	**0.0204**	**0.8572**	**0.1555**	**0.0078**	**53**
INR	**0.5236**	**0.4345**	**0.8403**	0.1510	**0.0033**	**54**	0.5245	0.4360	0.8392	**0.1503**	0.0034	52
JPY	**1.0706**	1.4721	0.6270	**0.3691**	**0.0066**	49	1.0770	**1.4700**	**0.6281**	0.3789	0.0066	**50**

TABLE 10.5
The Ten-Days-Ahead Prediction Performance of the LSTM and GRU Models

Currency	LSTM Model						GRU Model					
	MAPE	RMSE	R²	ARV	Theil's	DA	MAPE	RMSE	R²	ARV	Theil's	DA
AUD	1.4997	0.0148	0.3057	0.5800	0.0096	55	1.3444	0.0131	0.4567	0.5167	0.0085	56
CAD	1.2397	0.0198	0.7582	0.3019	0.0077	55	1.2609	0.0200	0.7535	0.2948	0.0077	55
GBP	1.5861	0.0269	0.7491	0.2889	0.0103	56	1.4124	0.0242	0.7969	0.2503	0.0093	57
INR	0.6116	0.5181	0.7673	0.2162	0.0040	49	0.5996	0.5001	0.7832	0.1985	0.0038	51
JPY	1.3903	1.8456	0.4078	0.5589	0.0083	51	1.2942	1.7395	0.4739	0.5396	0.0078	50

TABLE 10.6
The One-Day-Ahead Prediction Performance of the LSTM and GRU Models

Currency	LSTM Model						GRU Model					
	MAPE	RMSE	R²	ARV	Theil's	DA	MAPE	RMSE	R²	ARV	Theil's	DA
AUD	0.3934	0.0038	0.9540	0.0465	0.0025	57	0.3854	0.0038	0.9554	0.0449	0.0024	57
CAD	0.3614	0.0061	0.9769	0.0233	0.0024	59	0.3228	0.0055	0.9810	0.0194	0.0021	58
GBP	0.3983	0.0068	0.9841	0.0161	0.0026	54	0.4320	0.0073	0.9818	0.0186	0.0028	54
INR	0.1906	0.1706	0.9754	0.0245	0.0013	51	0.1901	0.1707	0.9754	0.0248	0.0013	49
JPY	0.4919	0.7028	0.9150	0.0802	0.0031	50	0.4183	0.6161	0.9347	0.0642	0.0028	51

TABLE 10.7

The Three-Days-Ahead Prediction Performance of the LSTM and GRU Models

Currency	LSTM Model						GRU Model					
	MAPE	RMSE	R²	ARV	Theil's	DA	MAPE	RMSE	R²	ARV	Theil's	DA
AUD	0.7919	0.0076	0.8153	0.1835	0.0050	55	0.7096	0.0068	0.8518	0.1488	0.0044	55
CAD	0.6863	0.0111	0.9233	0.0821	0.0043	58	0.7460	0.0119	0.9119	0.0915	0.0046	57
GBP	0.7442	0.0125	0.9460	0.0578	0.0048	54	0.7498	0.0126	0.9453	0.0581	0.0048	54
INR	0.3368	0.2894	0.9292	0.0695	0.0022	49	0.4399	0.3577	0.8918	0.1019	0.0028	50
JPY	0.8005	1.1159	0.7857	0.2062	0.0050	50	0.7520	1.0431	0.8127	0.1827	0.0047	51

TABLE 10.8

The Five-Days-Ahead Prediction Performance of the LSTM and GRU Models

Currency	LSTM Model						GRU Model					
	MAPE	RMSE	R²	ARV	Theil's	DA	MAPE	RMSE	R²	ARV	Theil's	DA
AUD	0.9427	0.0093	0.7248	0.2726	0.0060	59	0.9558	0.0094	0.7169	0.2874	0.0061	59
CAD	0.8518	0.0136	0.8859	0.1321	0.0053	56	0.8500	0.0136	0.8861	0.1257	0.0053	56
GBP	0.9937	0.0167	0.9032	0.1003	0.0064	58	0.9636	0.0162	0.9092	0.0994	0.0062	59
INR	0.4344	0.3658	0.8840	0.1119	0.0028	54	0.5691	0.4530	0.8221	0.1601	0.0035	53
JPY	0.9334	1.2612	0.7235	0.2802	0.0057	55	0.9145	1.2453	0.7304	0.2718	0.0056	54

TABLE 10.9
The Seven-Days-Ahead Prediction Performance of the LSTM and GRU Models

Currency	LSTM Model						GRU Model					
	MAPE	RMSE	R²	ARV	Theil's	DA	MAPE	RMSE	R²	ARV	Theil's	DA
AUD	1.1999	0.0117	0.5660	0.4089	0.0076	57	1.0788	0.0106	0.6406	0.3624	0.0069	56
CAD	1.0249	0.0164	0.8334	0.1984	0.0064	55	1.0200	0.0163	0.8356	0.1856	0.0063	55
GBP	1.2729	0.0215	0.8394	0.1672	0.0082	51	1.1683	0.0196	0.8660	0.1516	0.0075	52
INR	0.5267	0.4491	0.8252	0.1665	0.0035	54	0.5330	0.4405	0.8317	0.1564	0.0034	53
JPY	1.1341	1.5432	0.5860	0.4011	0.0069	49	1.1146	1.5135	0.6018	0.3987	0.0068	50

TABLE 10.10
The Ten-Days-Ahead Prediction Performance of the LSTM and GRU Models

Currency	LSTM Model						GRU Model					
	MAPE	RMSE	R²	ARV	Theil's	DA	MAPE	RMSE	R²	ARV	Theil's	DA
AUD	1.4841	0.0146	0.3102	0.5934	0.0095	55	1.3396	0.0130	0.4548	0.5253	0.0085	55
CAD	1.3119	0.0209	0.7305	0.3186	0.0081	55	1.2552	0.0199	0.7565	0.2934	0.0077	55
GBP	1.4583	0.0250	0.7812	0.2566	0.0096	56	1.4464	0.0248	0.7844	0.2587	0.0095	58
INR	0.5878	0.4900	0.7857	0.1988	0.0038	48	0.5888	0.4932	0.7829	0.1951	0.0038	51
JPY	1.2346	1.7001	0.4932	0.5443	0.0077	51	1.3514	1.8036	0.4296	0.5654	0.0081	50

TABLE 10.11
The One-Day-Ahead Prediction Performance of the LSTM and GRU Models

Currency	LSTM Model						GRU Model					
	MAPE	RMSE	R^2	ARV	Theil's	DA	MAPE	RMSE	R^2	ARV	Theil's	DA
AUD	0.3708	0.0037	0.9574	0.0421	0.0024	57	0.4139	0.0040	0.9497	0.0508	0.0026	57
CAD	0.3307	0.0056	0.9805	0.0199	0.0022	59	0.3275	0.0056	0.9807	0.0198	0.0022	58
GBP	0.3993	0.0068	0.9840	0.0161	0.0026	54	0.3958	0.0068	0.9843	0.0160	0.0026	55
INR	0.2427	0.2027	0.9653	0.0347	0.0016	51	0.2405	0.1952	0.9678	0.0318	0.0015	49
JPY	0.4156	0.6133	0.9353	0.0642	0.0028	50	0.5650	0.7871	0.8934	0.1005	0.0035	51

TABLE 10.12
The Three-Days-Ahead Prediction Performance of the LSTM and GRU Models

Currency	LSTM Model						GRU Model					
	MAPE	RMSE	R^2	ARV	Theil's	DA	MAPE	RMSE	R^2	ARV	Theil's	DA
AUD	0.7027	0.0068	0.8535	0.1469	0.0044	55	0.7280	0.0071	0.8413	0.1581	0.0046	55
CAD	0.6300	0.0103	0.9344	0.0716	0.0040	57	0.6600	0.0108	0.9281	0.0757	0.0042	56
GBP	0.7564	0.0127	0.9442	0.0567	0.0048	53	0.9254	0.0152	0.9202	0.0829	0.0058	54
INR	0.3974	0.3326	0.9041	0.0896	0.0026	50	0.3721	0.3145	0.9143	0.0822	0.0024	51
JPY	0.7336	1.0225	0.8182	0.1798	0.0046	51	0.7233	1.0027	0.8252	0.1756	0.0045	51

TABLE 10.13

The Five-Days-Ahead Prediction Performance of the LSTM and GRU Models

Currency	LSTM Model						GRU Model					
	MAPE	RMSE	R^2	ARV	Theil's	DA	MAPE	RMSE	R^2	ARV	Theil's	DA
AUD	**0.9351**	**0.0092**	**0.7297**	**0.2667**	**0.0060**	58	0.9847	0.0097	0.7005	0.2977	0.0063	59
CAD	**0.8515**	**0.0136**	**0.8856**	**0.1311**	**0.0053**	56	0.8814	0.0141	0.8780	0.1335	0.0054	56
GBP	**0.9469**	**0.0158**	**0.9131**	**0.0889**	**0.0060**	57	0.9548	0.0160	0.9109	0.1023	0.0061	59
INR	**0.4223**	**0.3581**	**0.8888**	**0.1109**	**0.0028**	**54**	0.4241	0.3609	0.8871	**0.1087**	**0.0028**	53
JPY	0.9701	1.3458	0.6851	0.2969	0.0060	**55**	**0.9111**	**1.2293**	**0.7373**	**0.2681**	**0.0055**	54

TABLE 10.14

The Seven-Days-Ahead Prediction Performance of the LSTM and GRU Models

Currency	LSTM Model						GRU Model					
	MAPE	RMSE	R^2	ARV	Theil's	DA	MAPE	RMSE	R^2	ARV	Theil's	DA
AUD	1.2148	0.0118	0.5485	0.4285	0.0077	57	**1.1356**	**0.0112**	**0.5996**	**0.3994**	**0.0073**	57
CAD	1.0414	0.0167	0.8291	0.2027	0.0064	55	**1.0168**	**0.0163**	**0.8367**	**0.1843**	**0.0063**	55
GBP	1.2382	0.0211	0.8446	0.1721	0.0081	52	**1.1744**	**0.0198**	**0.8629**	**0.1608**	**0.0076**	52
INR	**0.5220**	**0.4323**	**0.8332**	**0.1601**	**0.0033**	55	0.5243	0.4372	0.8294	0.1546	0.0034	52
JPY	**1.1153**	**1.5197**	**0.5950**	**0.3998**	**0.0068**	50	1.1595	1.5735	0.5659	0.4233	0.0071	49

TABLE 10.15
The Ten-Days-Ahead Prediction Performance of the LSTM and GRU Models

Currency	LSTM Model						GRU Model					
	MAPE	RMSE	R^2	ARV	Theil's	DA	MAPE	RMSE	R^2	ARV	Theil's	DA
AUD	1.3755	0.0134	0.4225	0.5294	0.0087	56	1.3174	0.0127	0.4779	0.5009	0.0083	55
CAD	1.4074	0.0225	0.6885	0.3523	0.0087	55	1.2817	0.0203	0.7465	0.3009	0.0078	55
GBP	1.8596	0.0307	0.6697	0.3690	0.0118	57	1.3180	0.0224	0.8241	0.1977	0.0086	57
INR	0.6131	0.5077	0.7699	0.2139	0.0039	48	0.5707	0.4841	0.7909	0.1868	0.0037	51
JPY	1.2529	1.7016	0.4923	0.5359	0.0076	49	1.2719	1.7172	0.4829	0.5405	0.0077	50

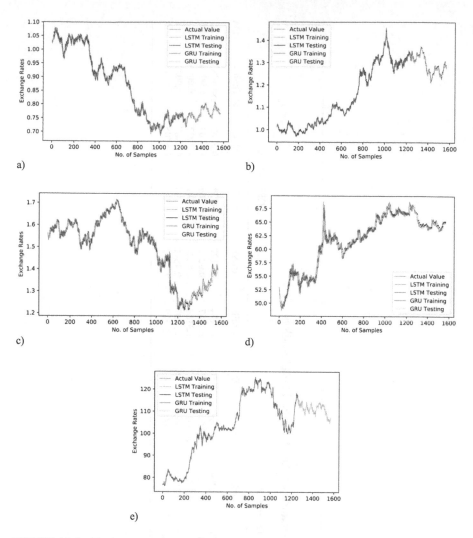

FIGURE 10.2 Exchange rate prediction graph of training and testing phase of LSTM model and GRU model for five different currencies with different sliding window sizes (WS): **(a)** five–days-ahead prediction of AUD with WS 10; **(b)** three-days-ahead prediction of CAD with WS 7; **(c)** seven-days-ahead prediction of GBP with WS 10; **(d)** ten-days-ahead prediction of INR with WS 13; and **(e)** one-day-ahead prediction of JPY for WS 7.

10.3.2 For Sliding Window Size 10

Tables 10.6–10.10 show the results for sliding window size 10. For AUD, CAD, and JPY daily forex, GRU gives better results; for the INR dataset, LSTM gives better results. When the samples have a long-term dependency problem, the LSTM model outperforms GRU, because it has more features and a more complex architecture than GRU.

10.3.3 FOR SLIDING WINDOW SIZE 13

From Tables 10.11–10.15 it is observed that LSTM and GRU give mixed results. LSTM is better for AUD, CAD, and JPY, when the number of the days-ahead prediction is less. The GRU model gives better results for INR, AUD, GBP, and CAD, when the number of the days-ahead prediction is more.

The simulation of exchange rate forecasting of five different currencies using LSTM and GRU models and the convergence characteristics of both models are given in Figures 10.2 and 10.3.

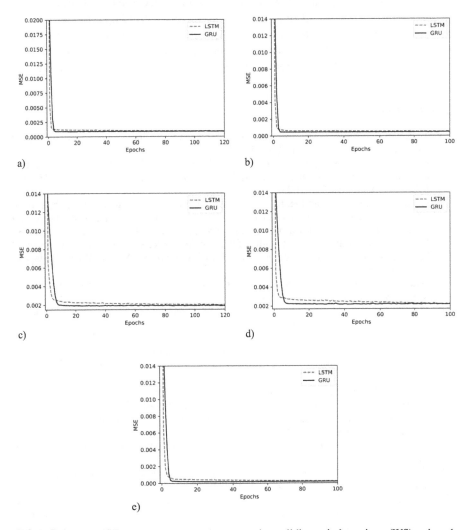

FIGURE 10.3 MSE convergence graph over various sliding window sizes (WS) using the LSTM and GRU model: **(a)** five-days-ahead prediction of AUD with WS 10; **(b)** three-days-ahead prediction of CAD with WS 7; **(c)** seven-days-ahead prediction of GBP with WS 10; **(d)** ten-days-ahead prediction of INR with WS 13; and **(e)** one-day-ahead prediction of JPY for WS 7.

10.4 CONCLUSION

This chapter has described the implementation of a special type of RNN, the GRU network for the exchange rate forecasting of five different currencies (AUD, CAD, GBP, INR, and JPY). The GRU model reduces the complex gate structure of LSTM to two gated architectures having update and reset gates, and it also decreases the number of iterations needed to execute the program. This leads to fewer in-memory calculations which makes the proposed model faster than LSTM. The performance measure of LSTM and GRU is carried out by applying different sliding window sizes for various days-ahead predictions. For sliding window size 7, the GRU model gives the best results for almost all five different currencies; for sliding window sizes 10 and 13, LSTM performs better in some cases, with the GRU results being better for the rest. Thus, by overall performance measure, GRU outperforms LSTM and converges faster than LSTM with greater accuracy and efficiency for windows size 7.

REFERENCES

1. Taylor, S. J. "Forecasting the volatility of currency exchange rates." *International Journal of Forecasting* 3.1 (1987): 159–170.
2. Kamruzzaman, J., R. A. Sarker, and I. Ahmad. "SVM based models for predicting foreign currency exchange rates." *Third IEEE International Conference on Data Mining*. IEEE, Los Alamitos, CA, 2003.
3. Nanda, S. K., R. Vyas, and H. K. Vamshidhar. "Forecasting foreign exchange rate using robust lagueree neural network." *2018 3rd International Conference and Workshops on Recent Advances and Innovations in Engineering (ICRAIE)*. IEEE, Jaipur, India, 2018.
4. Nagarajan, V., et al. "Forecast studies for financial markets using technical analysis." *2005 International Conference on Control and Automation*, Vol. 1. IEEE, Budapest, Hungary, 2005.
5. Galeshchuk, S. "Neural networks performance in exchange rate prediction." *Neurocomputing 172* (2016): 446–452.
6. Huang, W., et al. "Comparisons of the different frequencies of input data for neural networks in foreign exchange rates forecasting." In: *International Conference on Computational Science*. Springer, Berlin, 2006.
7. Bui, L. T. and T. T. H. Dinh. "A novel evolutionary multi-objective ensemble learning approach for forecasting currency exchange rates." *Data & Knowledge Engineering* 114 (2018): 40–66.
8. He, K., et al. "Forecasting exchange rate value at risk using deep belief network ensemble based approach." *Procedia Computer Science* 139 (2018): 25–32.
9. Das, S. R., D. Mishra, and M. Rout. "An optimized feature reduction based currency forecasting model exploring the online sequential extreme learning machine and krill herd strategies." *Physica A: Statistical Mechanics and Its Applications* 513 (2019): 339–370.
10. Yong, Y. L., et al. "Foreign currency exchange rate prediction using neuro-fuzzy systems." *Procedia Computer Science* 144 (2018): 232–238.
11. Andreou, A. S., E. F. Georgopoulos, and S. D. Likothanassis. "Exchange-rates forecasting: A hybrid algorithm based on genetically optimized adaptive neural networks." *Computational Economics* 20.3 (2002): 191–210.
12. Ni, H. and H. Yin. "Exchange rate prediction using hybrid neural networks and trading indicators." *Neurocomputing* 72.13–15 (2009): 2815–2823.
13. Jena, P. R., R. Majhi, and B. Majhi. "Development and performance evaluation of a novel knowledge guided artificial neural network (KGANN) model for exchange rate

prediction." *Journal of King Saud University-Computer and Information Sciences* 27.4 (2015): 450–457.

14. Rehman, M., G. M. Khan, and S. Ali Mahmud. "Foreign currency exchange rates prediction using cgp and recurrent neural network." *IERI Procedia 10* (2014): 239–244.

15. Ni, L., et al. "Forecasting of Forex time series data based on deep learning." *Procedia Computer Science* 147 (2019): 647–652.

16. Chao, J., F. Shen, and J. Zhao. "Forecasting exchange rate with deep belief networks." *The 2011 International Joint Conference on Neural Networks.* IEEE, San Jose, CA, 2011.

17. Shen, F., J. Chao, and J. Zhao. "Forecasting exchange rate using deep belief networks and conjugate gradient method." *Neurocomputing* 167 (2015): 243–253.

18. Naik, N. and B. R. Mohan. "Stock price movements classification using machine and deep learning techniques—The case study of Indian Stock Market." In: *International Conference on Engineering Applications of Neural Networks.* Springer, Cham, 2019.

19. Yu, L., S. Wang, and K. K. Lai. "Adaptive smoothing neural networks in foreign exchange rate forecasting." In: *International Conference on Computational Science.* Springer, Berlin, Heidelberg, 2005.

20. Hussain, A. J., et al. "Regularized dynamic self-organized neural network inspired by the immune algorithm for financial time series prediction." *Neurocomputing* 188 (2016): 23–30.

21. Usmani, M., et al. "Predicting market performance with hybrid model." *2018 3rd International Conference on Emerging Trends in Engineering, Sciences and Technology (ICEEST).* IEEE, Karachi, Pakistan, 2018.

22. Kamruzzaman, J. and R. A. Sarker. "Forecasting of currency exchange rates using ANN: A case study." *International Conference on Neural Networks and Signal Processing, 2003,* Vol. 1. IEEE, Nanjing, 2003.

23. Majhi, R., G. Panda, and G. Sahoo. "Efficient prediction of foreign exchange rate using nonlinear single layer artificial neural model." *2006 IEEE Conference on Cybernetics and Intelligent Systems.* IEEE, Bangkok, Thailand, 2006.

24. He, H. and X. Shen. "Bootstrap methods for foreign currency exchange rates prediction." *2007 International Joint Conference on Neural Networks.* IEEE, Atlanta, GA, 2007.

25. Ince, H. and T. B. Trafalis. "A hybrid model for exchange rate prediction." *Decision Support Systems* 42.2 (2006): 1054–1062.

26. Zhang, K., et al. "Stock market prediction based on generative adversarial network." *Procedia Computer Science* 147 (2019): 400–406.

27. Ranjit, S., et al. "Comparison of algorithms in Foreign Exchange Rate Prediction." *2018 IEEE 3rd International Conference on Computing, Communication and Security (ICCCS).* IEEE, Kathmandu, Nepal, 2018.

28. Naik, N. and B. R. Mohan. "Study of stock return predictions using recurrent neural networks with LSTM." In: *International Conference on Engineering Applications of Neural Networks.* Springer, Cham, 2019.

29. Chung, J., et al. "Empirical evaluation of gated recurrent neural networks on sequence modeling." arXiv preprint arXiv:1412.3555 (2014).

30. Hochreiter, S. and J. Schmidhuber. "Long short-term memory." *Neural Computation* 9.8 (1997): 1735–1780.

11 Optimal Selection of Features Using Teaching-Learning-Based Optimization Algorithm for Classification

Himansu Das, Soham Chakraborty, Biswaranjan Acharya, and Abhaya Kumar Sahoo

CONTENTS

11.1 INTRODUCTION

In real-life scenarios, we often deal with large amounts of data [1], where most of the features are not relevant for prediction. These features or attributes have a huge influence on effectiveness to determine the performance of the model. These irrelevant or redundant features can also negatively impact on the performance of the model. So, we need very refined data, with non-redundant and relevant features only. To address this issue, feature selection (FS) [2] plays a decisive role in the selection of relevant features. In machine learning (ML) and statistics, FS is also known as feature or attribute or variable selection. It is the process in which the subset of relevant features is selected for the construction of the model. It is also the key concept of ML in which performance can hugely impact on the classification [3–10] models, and it is the method in which the automatic selection of features is made that contributes the most to the prediction variable in which we are interested [11]. The use of FS techniques reduces the overfitting issue, improves classification accuracy, and also reduces computational time. Similarly, feature reduction [12–14] is another form of the dimensionality reduction approach, which seeks to reduce the number of attributes.

In the feature reduction method, new combinations of features are constructed. However, FS methods can either include or exclude the features without altering its significance. The selected significant features consist of an optimized feature representation of the original features. Indeed, FS has been shown to be a critical task for enhancing the decision-making process in various domains. This research proposes an FS approach, based on the teaching-learning-based optimization (TLBO) [15, 16] algorithm, along with the evidence of its superiority over other models.

FS is an NP-hard problem [17] as it takes exponential time to find the optimal features. In the traditional search approach, the possible solution space increases exponentially with an increase in the number of features to find the optimal feature set. Hence, it is essential to reduce the dimension of the features that will increase the performance of the model and also decrease the computational cost of the model. This motivated us to design a suitable FS approach that selects an optimal subset of features and also reduces the dimensionality of the problem. It simply filters the most irrelevant features to improve the performance of the classification [18–20] model and also reduces the computational cost of the problem. The several wrapper-based FS approaches using a genetic algorithm (GA) [21] have been used to solve FS problems [22, 23]. The GA-based FS approach needs several parameters to be tuned to obtain better performance. Much other work has been done using differential evolution (DE) [24] to solve FS problems [25, 26]. The DE-based FS approach is a more complex, large, and multi-modal search activity, for which the model will be more complex. Particle swarm optimization (PSO) [27] has been extensively used in the FS [28, 29] process. This PSO-based FS approach suffers from premature convergence regarding the local optimum as particles learn from the personal best and the global best. These optimization algorithms have some controlling parameters, such as population size, and the number of generations need to be tuned to increase the performance of the optimization algorithm. Apart from these two controlling parameters, every other optimization algorithm has some algorithm-specific controlling parameters, such as crossover probability, mutation probability, and a selection operator in the case of GA, with DE also having crossover and mutation parameters; similarly, in PSO, social and cognitive parameters, and initial weights are also to be properly tuned to get the optimal result. The proper tuning of these parameters affects the performance of the model. However, improper tuning of these algorithm-specific parameters leads to an increase in the computational cost and also traps in the local optimum. Rao et al. developed a TLBO optimization algorithm that does not depend on any algorithmic-specific controlling parameters. To address these issues, we have proposed an FS approach based on the TLBO optimization algorithm named FSTLBO to find the best suitable optimal features in order to improve the performance of the model. This approach addresses the problems associated with the existing approaches of FS and finds the best suitable relevant features, by updating the worst features, that take on the strength of the TLBO optimization algorithm. These selected features are passed to the learning model to compute the fitness (error) and accordingly update the position of individuals based on the fitness. Here, we perform the FS by using well-known classification algorithms called K-nearest neighbors (KNN). This classification algorithm is wrapped with four optimization techniques (GA, DE, PSO, and TLBO) to form four FS approaches, namely FSGA, FSDE,

FSPSO, and FSTLBO. It is observed that the FSTLBO method takes the advantages of non-algorithm-specific parameters over other algorithm-specific parameters.

The rest of this chapter is as follows. Section 11.2 presents related work based on FS and optimization-based feature-subset selection and its implementation. Section 11.3 describes the basic technologies where the tools and techniques used in this research are addressed. Section 11.4 addresses the proposed model in which we have diagrammatically represented the flow model of our work. Section 11.5 discusses the analysis of the results of several FS approaches and the implementation methodologies in depth. Section 11.6 concludes with future scope of the work.

11.2 RELATED WORK

In this section, we discuss FS techniques based on several optimization algorithms. FS was there among the foremost techniques of researchers for many decades. A survey was conducted by Dash and Liu on several FS techniques [30] and their procedures. One of the recent studies done by Xue et al. [31] considers the broad evaluation of FS problems, which was done based on several evolutionary techniques, analyzing each performance with their individual objectives. The different advantages and challenges of various FS algorithms are also addressed. It has also been stated that by eliminating the number of features the accuracy of the classification or regression model can be much improved. This also addresses various algorithms that have been used to solve problems on FS. Yang and Honavar [32] proposed an algorithm that was based on the combination of a genetic algorithm and a neural network model for obtaining a suitable subset of features for classification. The tests were evaluated on several benchmark datasets to display the progress of the results obtained from the model by using all features. Inza et al. [33] gave a detailed description of FS problems via estimation of a Bayesian network algorithm. When there is limited data about the domain, a randomized or evolutionary search algorithm can be applied, which has been derived from the distribution algorithms. The Naive Bayes and ID3 algorithms have been used for the experiment; as a result FS does not have a significant influence over the accuracy; however, it minimizes the CPU execution time dramatically. Huang and Wang [34] proposed a model on a genetic algorithm, optimized the process of FS, and tuned the SVM parameters. They made a comparison of their model with the greedy algorithms that are mostly used for parameter searching. Experiments have been done using 11 known real-world datasets to present the approach that has a significant effect on the accuracy of the classification model positively. Cervante et al. [35] produced a combination of PSO with two information metrics, entropy and mutual information. In classification problems, they applied decision trees.

11.3 BASIC TECHNOLOGY

There are different FS techniques, such as filter, wrapper, and embedded methods. We are mostly concerned to discuss the wrapper methods. The FS model in the wrapper class is mainly based on the ML algorithms that are to be trained. It evaluates all possible combinations of features against the evaluation criterion. In the background, it basically follows a greedy algorithm. The evaluation criterion mentioned earlier is fitness, but the performance measure depends on the type of problem it is dealing with. Some of the regression evaluation criteria are p-values, R-squared, and adjusted R-squared. Some of the classification evaluation criteria are accuracy, precision, recall, and F1 score. Finally, the most appropriate combination will be selected for the useful features. This combination of features gives the optimal results for the ML algorithm it has been applied to.

Some of the techniques under wrapper methods that are most commonly used are (1) forward selection, (2) backward selection, and (3) bi-directional elimination (stepwise selection). We emphasize the above-mentioned techniques under wrapper methods. Basically, forward selection begins with an intercept. The testing of the various variables is performed for checking relevance. The "best" variable, where "best" is determined by some predefined criteria, is added to the model. We continue the process, adding variables one by one and testing at each step while the model continues to improve. Once the model stops improving then we stop adding any further variables. The criteria on which we determine whether to consider a variable or not varies. It may be finding the lowest score in cross-validation, the lowest p-value, or any other tests or measures of accuracy. Unlike forward selection, backward selection begins with a full least squares model that contains all predictors, and then, gradually in an iterative manner, removes the least useful predictor, one by one. Backward selection is done in those problems where we have more observations than variables because we can do least squares regressions when the number of variables is more than the number of features. If the opposite situation arises then we won't be able to fit a least squares model. Bidirectional elimination is necessarily a forward selection technique, but it has the possibility of detecting a selected variable at each and every stage, as in backward elimination, whenever correlation is found between the variables. It is mostly used as the default approach. Feature subset selection is defined as a process in which we choose some selective features from the whole feature space. There are many diverse concepts on FS approaches in the literature. Some mainly deal with the size of the selected data, while other concepts are all about improving prediction accuracy. Essentially FS constructs a dataset which most effectively represents the original data. The main objective is to just find the minimum number of features, maintaining a higher end accuracy in the classification process. Our research work surrounds the same discussed area and we have mainly dealt with overcoming the challenges faced for choosing the most suitable subset of features among the whole feature space.

11.4 PROPOSED MODEL

In this section, the working model of the FSTLBO approach is presented with a detailed explanation. In this approach, the algorithm starts by random initial population generation (both for the teacher and students). The population is the set of binary solutions, which helps to select or reject the features, based on the value 1 or 0. Here, we have inculcated this teaching-learning-based optimization technique along with the concept of a wrapper FS technique. The basic underlying concept in wrapper-based feature-subset selection is that it randomly selects a set of features among all the features in the original dataset. Over several iterations, it chooses different sets and evaluates the fitness for individual selection. The one with the maximum fitness is chosen and selected as the feature. There are three types of wrapper-class selection techniques, forward selection, backward selection, and bidirectional elimination. After implementing the wrapper-class technique along with teaching-learning optimization, performance is evaluated. The overall methodology includes the following steps: (1) the data is collected from different sources; (2) this raw data contains a lot of inconsistencies, and irrelevant and null values; (3) the redundant data are removed by FS methods. Here we have used the wrapper-selection method, henceforth converting the higher dimensional data to lower dimensional data and removing unnecessary information. The mathematical reason behind this is to extract only the significant features that have the maximum correlation with the prediction variable. (4) Then we apply our classification model and evaluate the accuracy and precision value.

Figure 11.1 shows the self-explanatory working model of the FSTLBO model for classification. In this model, the dataset is categorized into training and testing data. Then, a random binary population of size 10 is generated. The binary value 1

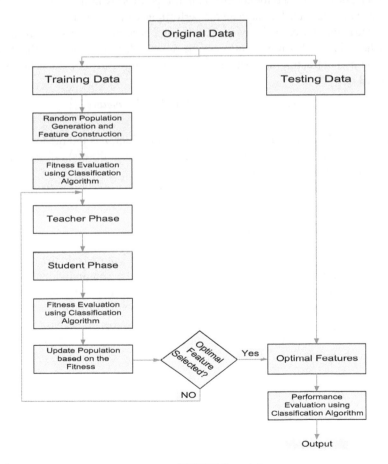

FIGURE 11.1 Work-flow of the proposed FSTLBO model.

indicates the selection of a feature and 0 indicates the rejection of a feature. By combining the training data with the binary population, the new feature set is constructed. This newly constructed feature set is passed to the classification algorithm for fitness evaluation. The newly constructed features are also passed through the teacher and student phase to obtain the updated population. The fitness of the updated population is computed using a classification algorithm and comparing with the old fitness, which accordingly updates the population and fitness. This process is repeated for certain iterations to acquire the optimal features, which are passed to the classification algorithms for performance computation.

The detailed working principle of the FSTLBO method for finding the optimal features will now be described. This is a two-phase process, such as teacher and student, in which the teacher trains the students. Let us consider the dataset $X = [X_{i,1}, X_{i,2}, X_{i,3}, ..., X_{i,D}]$, where D is the dimensions (subjects) of the problem and $i = 1, 2, 3, ...,$ m is the number of instances. Initially, a random binary population is generated and combined with the training data to generate the new feature set. The fitness (error) of the newly constructed feature set is computed by using a classification algorithm. Here, the KNN algorithm is used for computation of the fitness with the value of $k = 5$. This fitness is to be minimized to increase the performance of the model.

In the teacher phase, the individual having the minimum fitness will be treated as the teacher. The teacher tries to increase the mean value of the subject that he or she has taught by subtracting the mean result of each subject from the teacher by using Equation 11.1.

$$Difference(j) = r \times (teacher(j) - TF \times mean(j)) \quad (11.1)$$

Here, $j = 1, 2, 3, ..., D$ is the dimension of the problem, r is the random number in the range [0–1], $teacher$ is the best individual (minimum fitness), $mean$ is the mean of each subject, and TF is the teaching factor.

The new population can be generated by using Equation 11.2, where $k = 1, 2, 3, ..., N$ is the number of the population.

$$X_{New}(k) = X(k) + Difference \quad (11.2)$$

The transfer function is used to convert the continuous features into their corresponding binary features by computing the probability of the new individuals X_{New} using Equation 11.3. The population of each individual is updated by using Equation 11.4. The new individuals are flipped to 1 only if the probability of the new individual $P(X_{New}(k,j))$ is greater than or equal to the random number; otherwise it is set to 0, where $rand()$ is the random number distributed in the range of 0 to 1, and P represents the probability.

$$P(X_{New}(k,j)) = \frac{1}{(1 + \exp(-10 \times (X_{New}(k,j) - 0.5)))} \quad (11.3)$$

$$X_{New}(k,j) = \begin{cases} 1, & \text{if } P(X_{New}(k,j)) > rand() \\ 0, & \text{Otherwise} \end{cases} \quad (11.4)$$

The fitness of the newly generated population is computed and compared with the old population accordingly; the population and the fitness are updated based on the fitness value (minimum).

In the student phase, students try to increase their performance by interacting among themselves. Any one student interacts with any other student randomly to update his or her knowledge. This can be achieved by generating two unique random integer numbers $r1$ and $r2$ within the range of the population size N. The knowledge of the student is updated by using Equation 11.5.

$$X_{Student}(k) = \begin{cases} X_{New}(k) + rand(\ ) \times (X_{New}(r1) - X_{New}(r2)), & fitness(r1) \geq fitness(r2) \\ X_{New}(k) + rand(\ ) \times (X_{New}(r2) - X_{New}(r1)), & \text{Otherwise} \end{cases} \quad (11.5)$$

Again, the transfer function is applied to convert the continuous features into their corresponding binary features by computing the probability of the student individuals $X_{Student}$ using Equation 11.6. The population of each individual is updated by using Equation 11.7. The new individuals are flipped to 1 only if the probability of student individual $P(X_{Student}(k, j))$ is greater than or equal to the random number; otherwise it is set to 0, where $rand()$ is the random number distributed in the range of 0 and 1, and P represents the probability.

$$P\left(X_{Student}(k,j)\right) = \frac{1}{\left(1 + \exp\left(-10 \times \left(X_{Student}(k,j) - 0.5\right)\right)\right)} \quad (11.6)$$

$$X_{Student}(k,j) = \begin{cases} 1, & \text{if } P\left(X_{Student}(k,j)\right) > rand(\ ) \\ 0, & \text{Otherwise} \end{cases} \quad (11.7)$$

The fitness of the newly generated student population is computed and compared with the old population accordingly; the population and the fitness are updated based on the fitness value (minimum). This process repeats for certain iterations, or until the stopping criteria are reached, in order to obtain the optimal feature set, which is finally passed to the classification algorithm for performance evaluation.

11.5 RESULT ANALYSIS

This section discusses the detailed result analysis of the FSTLBO approach. Initially, seven datasets are collected from different online repositories, such as the UCI ML Repository [36]. These datasets were quite large in size with a large number of features. The main objective is to reduce the size of the datasets and their redundant features, which can decrease the efficiency of the predictions. We need to reduce the redundancy and the size of the dataset so that the cost of computation reduces, as well as the performance improving. Firstly, the raw data is pre-processed (for cleaning and transforming the data) in order to make it useful. Here, the missing values are also handled by imputing the median. The reason behind this is that there may be some outliers in the dataset. So, the average or mean could be erroneous and thus we have taken the median, which is basically the center value of the whole population; the outliers don't affect them. As there are a variety of parameters for every observation, so their ranges may vary, gradually leading to poor evaluation of predictions. To handle this problem, min-max normalization is used to scale the data in the range of 0.0–1.0. This helps to maintain the uniformity of the overall dataset, otherwise it would be much more difficult to visualize or represent all of the data practically and the models also would not perform well comparatively. We selected the most significant features by considering the fact that the data set may contain some features that may have become redundant in nature and, hence, in turn, may reduce the efficiency of the model used. Before applying the classification algorithms to the raw data, we selected the optimal features that have the maximum relevance or correlation with the prediction variable. In datasets, usually while collecting the data, more than one doctor performs different tests and collects the results. There may be a chance of redundant tests or similar kinds of tests by two or more doctors. These features are added to the feature space of the original dataset and hence the underlying distribution of the data becomes distorted. So, to improve the distribution, by reducing the features and choosing the feature subset that is the fittest among all the features in the original dataset. Here, we developed FS techniques using several optimization techniques. These were experimented with and evaluated to further improve the accuracy of the model. The whole dataset was cross-validated with a ten-fold process. Many times, due to the non-uniform distribution of the data points, the accuracy varied a lot. For evaluation of the final accuracy, we ran all the models ten times and considered their mean values. So, if we iterate over several accuracies and find the mean of them, then a much better and appropriate result can be obtained and hence the issue is handled. After applying the classification algorithms, we evaluated the accuracy in each model for all the datasets and concluded that our proposed FSTLBO model performed better compared to existing models.

In this experiment, we used a computer with an intel i5 processor with 8 GB of main memory. Matlab 2015a was used to implement all the FS and classification models. This experiment considered the random binary populations of size 10 and the number of iteration was 100. The higher the values of these two parameters, the higher the performance; but the execution time was high. The accuracy results obtained for seven datasets with the four FS models FSGA, FSDE, FSPSO, and FSTLBO are given in Table 11.1. The highest accuracy and optimum features are presented in bold type for seven of the datasets of the four models. These FS models used KNN as the classification algorithm to evaluate the performance with the k value as 5. The result reveals that the proposed FSTLBO model is superior in accuracy and in the minimal number of selected features as compared to existing models. The error plot of all the datasets are shown in Figure 11.2. The plots clearly indicate that the error of the FSTLBO model is less compared to the other existing models of FSGA, FSDE, and FSPSO.

TABLE 11.1

Accuracy of FS Models with KNN Classifier (%)

					KNN			
	FSGA	# of Features Selected Using FSGA	FSDE	# of Features Selected Using FSDE	FSPSO	# of Features Selected Using FSPSO	FSTLBO	# of Features Selected Using FSTLBO
Arrhythmia	65.01804	**130**	62.03328	217.4	63.15375	136.7	**66.9593**	135.7
Ionosphere	86.63831	13.2	84.61769	18.6	86.89642	11.8	**90.36983**	**9.1**
Pima	74.79221	5	74.06444	5.9	74.71095	4.9	**75.11723**	**4.6**
Colon	87.35713	993.3	86.40477	1531	87.83335	1006.7	**88.04763**	**992.1**
SPECTF heart	85.35716	21.5	83.35125	32.7	84.23782	**19**	**85.05042**	20.5
Sonar	86.25952	**27.1**	84.71428	43.1	86.07618	29.8	**86.85237**	28.5
Australian	84.73913	4.2	78.55072	7.7	**86.69566**	**3.9**	86.28984	**3.9**

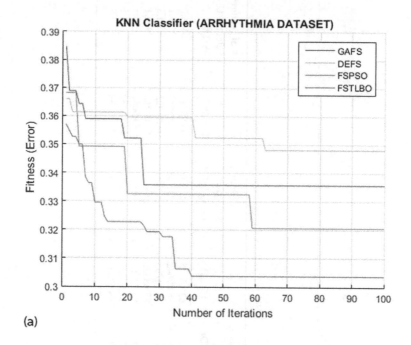

(a)

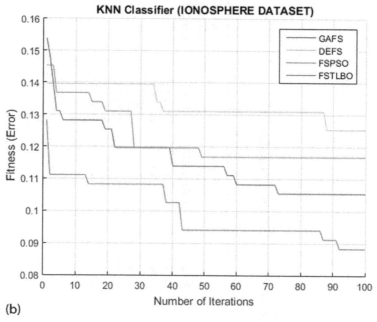

(b)

FIGURE 11.2   Error plots of FS approaches using KNN classifier for (a) Arrhythmia Dataset; (b) Ionosphere Dataset

(*Continued*)

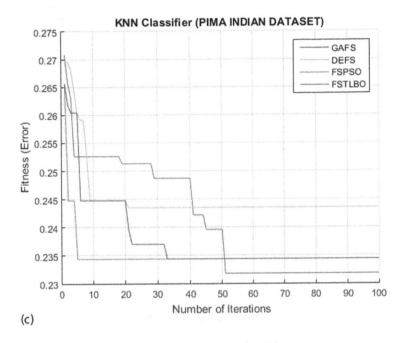

(c)

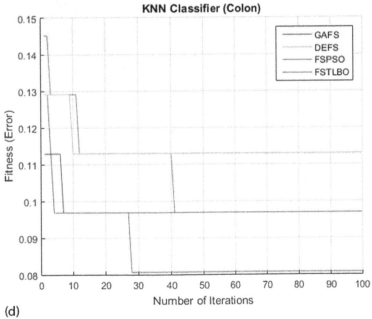

(d)

FIGURE 11.2 (CONTINUED)   (c) Pima Indian Dataset; (d) Colon Wisconsin Diagnosis Dataset

(Continued)

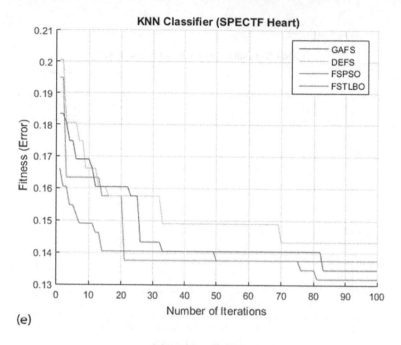

(e)

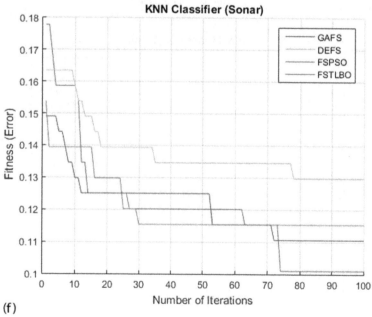

(f)

FIGURE 11.2 (CONTINUED)   (e) SPECTF Dataset; (f) Sonar Dataset

(*Continued*)

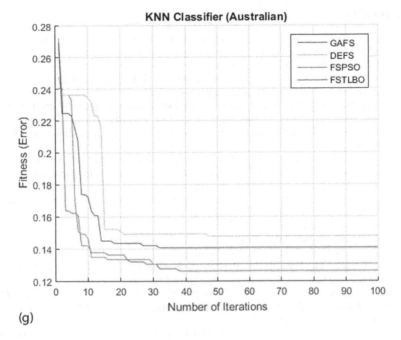

(g)

FIGURE 11.2 (CONTINUED)   (g) Australian Dataset.

11.6 CONCLUSION

The main objective behind this research was to reduce the redundancy in features so that the feature space is reduced from that of the size of the original dataset. To resolve this issue, several FS techniques were used, which take off only a subset of the overall feature space according to its significance and compute the minimal subset of features that provide a higher classification accuracy. We incorporated the idea of FS along with several optimization techniques for optimal FS. This further improves the accuracy while reducing the feature space, by using an optimal selection of the subset features. Mainly, we focused on the TLBO algorithm for optimal FS along with the KNN classification algorithm. Among all the FS models, the FSTLBO provides a better result, in terms of the optimal number of selection of features and accuracy as compared to other models. In the future, a multi-objective TLBO FS could be designed with several other datasets and the result compared with some other existing FS models. Moreover, some other classification algorithms could be applied in the classification phase to test the efficiency of the FS models.

REFERENCES

[1] Manyika, J., Chui, M., Brown, B., Bughin, J., Dobbs, R., Roxburgh, C., & Byers, A. H. (2011). *Bigdata: The Next Frontier for Innovation, Competition, and Productivity.* McKinsey Global Institute, New York.

[2] Li, J., Cheng, K., Wang, S., Morstatter, F., Trevino, R. P., Tang, J., & Liu, H. (2017). FS: A data perspective. *ACM Computing Surveys (CSUR)*, 50(6), 1–45.

[3] Das, H., Naik, B., & Behera, H. S. (2020). An experimental analysis of machine learning classification algorithms on biomedical data. In *Proceedings of the 2nd International Conference on Communication, Devices and Computing* (pp. 525–539). Springer, Singapore.

[4] Das, H., Naik, B., & Behera, H. S. (2018). Classification of diabetes mellitus disease (DMD): A data mining (DM) approach. In *Progress in Computing, Analytics and Networking* (pp. 539–549). Springer, Singapore.

[5] Das, H., Jena, A. K., Nayak, J., Naik, B., & Behera, H. S. (2015). A novel PSO based back propagation learning-MLP (PSO-BP-MLP) for classification. In *Computational Intelligence in Data Mining* (Vol. 2, pp. 461–471). Springer, New Delhi.

[6] Das, H., Dey, N., & Balas, V. E. (Eds.). (2019). *Real-Time Data Analytics for Large Scale Sensor Data.* Academic Press, London.

[7] Dey, N., Das, H., Naik, B., & Behera, H. S. (Eds.). (2019). *Big Data Analytics for Intelligent Healthcare Management.* Academic Press, Boston, MA.

[8] Dey, N., Ashour, A. S., Kalia, H., Goswami, R., & Das, H. (2019). *Histopathological Image Analysis in Medical Decision Making* (pp. 1–340). IGI Global, Hershey, PA. doi:10.4018/978-1-5225-6316-7

[9] Sahoo, A. K., Mallik, S., Pradhan, C., Mishra, B. S. P., Barik, R. K., & Das, H. (2019). Intelligence-based health recommendation system using big data analytics. In *Big Data Analytics for Intelligent Healthcare Management* (pp. 227–246). Academic Press, Boston, MA.

[10] Pradhan, C., Das, H., Naik, B., & Dey, N. (Eds.). (2018). *Handbook of Research on Information Security in Biomedical Signal Processing.* IGI Global, Hershey, PA.

[11] Kumar, V. & Minz, S. (2014). FS: A literature review. *SmartCR*, 4(3), 211–229.

[12] Das, H., Naik, B., Behera, H. S., Jaiswal, S., Mahato, P., & Rout, M. (2020). Biomedical data analysis using neuro-fuzzy model with post-feature reduction. *Journal of King Saud University-Computer and Information Sciences.*

[13] Das, H., Naik, B., & Behera, H. S. (2020). A hybrid neuro-fuzzy and feature reduction model for classification. *Advances in Fuzzy Systems*, 2020, 1–15.

[14] Das, H., Naik, B., & Behera, H. S. (2020). Medical disease analysis using neuro-fuzzy with feature extraction model for classification. *Informatics in Medicine Unlocked*, 18, 100288.

[15] Rao, R. V., Savsani, V. J., & Vakharia, D. P. (2011). Teaching–learning-based optimization: A novel method for constrained mechanical design optimization problems. *Computer-Aided Design*, 43(3), 303–315.

[16] Rao, R. (2016). Review of applications of TLBO algorithm and a tutorial for beginners to solve the unconstrained and constrained optimization problems. *Decision Science Letters*, 5(1), 1–30.

[17] Liu, H., & Yu, L. (2005). Toward integrating feature selection algorithms for classification and clustering. *IEEE Transactions on Knowledge & Data Engineering*, 17(4), 491–502.

[18] Das, H., Naik, B., & Behera, H. S. (2020). Disease classification using linguistic neuro-fuzzy model. In *Progress in Computing, Analytics and Networking* (pp. 45–53). Springer, Singapore.

[19] Rout, J. K., Rout, M., & Das, H. (2020). *Machine Learning for Intelligent Decision Science.* Springer, Singapore

[20] Sahoo, A. K., Pradhan, C., & Das, H. (2020). Performance evaluation of different machine learning methods and deep-learning based convolutional neural network for health decision making. In *Nature Inspired Computing for Data Science* (pp. 201–212). Springer, Cham.

[21] Holland, J. H. (1992). *Adaptation in Natural and Artificial Systems: An Introductory Analysis with Applications to Biology, Control, and Artificial Intelligence.* MIT Press, Cambridge, MA.

[22] Huang, J., Cai, Y., & Xu, X. (2007). A hybrid genetic algorithm for feature selection wrapper based on mutual information. *Pattern Recognition Letters*, 28(13), 1825–1844.

[23] Tan, F., Fu, X., Zhang, Y., & Bourgeois, A. G. (2008). A genetic algorithm-based method for feature subset selection. *Soft Computing*, 12(2), 111–120.

[24] Storn, R., & Price, K. (1997). Differential evolution—A simple and efficient heuristic for global optimization over continuous spaces. *Journal of Global Optimization*, 11(4), 341–359.

[25] Khushaba, R. N., Al-Ani, A., & Al-Jumaily, A. (2008, December). Differential evolution based feature subset selection. In *2008 19th International Conference on Pattern Recognition* (pp. 1–4). IEEE, Piscataway, NJ.

[26] Ghosh, A., Datta, A., & Ghosh, S. (2013). Self-adaptive differential evolution for feature selection in hyperspectral image data. *Applied Soft Computing*, 13(4), 1969–1977.

[27] Kennedy, J., & Eberhart, R. (1995, November). Particle swarm optimization. In *Proceedings of ICNN'95-International Conference on Neural Networks* (Vol. 4, pp. 1942–1948). IEEE, New York.

[28] Sharkawy, R. M., Ibrahim, K., Salama, M. M. A., & Bartnikas, R. (2011). Particle swarm optimization feature selection for the classification of conducting particles in transformer oil. *IEEE Transactions on Dielectrics and Electrical Insulation*, 18(6), 1897–1907.

[29] Sakri, S. B., Rashid, N. B. A., & Zain, Z. M. (2018). Particle swarm optimization feature selection for breast cancer recurrence prediction. *IEEE Access*, 6, 29637–29647.

[30] Dash, M., & Liu, H. (1997). FS for classification. *Intellgent Data Analysis* 1(3), 131–156.

[31] Xue, B., Zhang, M., Browne, W., & Yao, X. (2016). A survey on evolutionary computation approaches to FS. *IEEE Transactions of Evolutionary Computation*, 20(4), 606–626.

[32] Yang, J., & Honavar, V. (1998). Feature subset selection using a genetic algorithm. In *Feature Extraction, Construction and Selection* (pp. 117–136). Springer, New York.

[33] Inza, I., Larrañaga, P., Etxeberria, R., & Sierra, B. (2000). Feature subset selection by Bayesian network-based optimization. *Artificial Intelligence* 123(1), 157–184.

[34] Huang, C.-L., & Wang, C.-J. (2006). A GA based FS and parameters optimization for support vector machines. *Expert Systems with Applications*, 31(2), 231–240.

[35] Cervante, L., Xue, B., Zhang, M., & Shang, L. 2012. Binary particle swarm optimisation for FS: A fifilter based approach. In *Proceedings of the IEEE Congress on Evolutionary Computation (CEC)*. IEEE, Brisbane, Australia, pp. 1–8.

[36] *UCI ML Repository*. http://archive.ics.uci.edu/ml. University of California, School of Information and Computer Science, Irvine, CA.

12 An Enhanced Image Dehazing Procedure Using CLAHE and a Guided Filter

Badal Soni and Ujwala Baruah

CONTENTS

12.1 INTRODUCTION

The presence of unwanted particles in the atmosphere, such as water droplets, fog, and haze [24], makes it very difficult to get a clear image in an outdoor scene. The image becomes unclear because of the absorption and scattering of lights by these atmospheric particles. As a result, sometimes the contrast of images taken outside becomes reduced and their visibility may also be affected. This, in turn, can cause a major problem in different areas like photography, surveillance, navigation, and monitoring systems. Therefore, some measures to remove fog or haze have become an important concern in commercial as well as scientific applications. Here, we are focusing on haze, which occurs due to the presence of any particle along with water droplets in the atmosphere. Haze is mostly found in areas having high pollution.

It can generally be described as the union of airlight and attenuation, where airlight occurs when light gets scattered and reflected back in the direction of the source, which causes the whiteness effect in the image, and where attenuation is the diminishing effect of the image contrast.

The following model is used widely to describe hazy image formation [7, 9, 25, 35]:

$$I(x) = J(x) \times t(x) + A \times (1 - t(x)) \tag{12.1}$$

where the input hazy image is expressed as $I(x)$, $t(x)$ is the transparency of haze ranging from 0 to 1, which works by blending atmospheric light (A) and the original color of the image ($J(x)$). The $J(x) \times t(x)$ is called attenuation and $A \times (1 - t(x))$ is called as airlight. The transmission factor $t(x)$ is decreased with increasing depth of the object, which is expressed as:

$$t(x) = e^{-\beta d(x)} \tag{12.2}$$

where β stands for the scattering coefficient of the atmosphere and $d(x)$ is the scene depth.

The technique which is used to recover the original image from the hazy image is called a dehazing technique. The conventional technique to remove haze involves mainly two steps: first to remove the airlight and second to recover the lost information from the image. Basically, dehazing methods can be classified into two categories: Multiple image-based dehazing methods and single image-based dehazing methods. Some relevant work based on machine learning and image processing are described in [4, 31, 40].

Multiple image-based methods require multiple images of the same scene to execute, whereas single image-based methods need only one image to carry out the process.

Multiple image-based methods can again be extended into the following categories: A weather-condition-based method [21, 25] which depends on the variation of the same image taken during different weather conditions. Though this method can enhance the image visibility it cannot work for dynamic images. For the polarization-based method [29, 30], multiple images having a different filter of polarization are considered. This polarization filter can be adjusted by the attached polarization filter present in the camera. The limitation with this method is that it cannot work properly when the haze is thick or dense. Similarly for the depth map-based method [14, 23], some depth information [29, 30] is used to dehaze an image. This works on a 3D model, and the appearance of the input scene gives depth and in turn helps in removing haze from the image. This technique needs user interaction, that is, it is not automatic.

To cope with the limitation of the multiple image-based method, single image-based methods are used which can also be sub-categorized into the following. The Contrast Maximization method [35], which helps in enhancing the contrast of the input hazy image [25]. The output images will have a large value of saturation as the proposed method only enhances the visibility but does not improve the overall brightness of the image. The method of Independent Component Analysis [8], relies on statistics dividing two additive components from an input hazy image [21]. This approach achieves good results. However, in the presence of dense haze the performance gets degraded.

Therefore, Dark Channel Prior (DCP) [10, 22, 39] is introduced, where it is assumed that some of the image pixels have very less intensity in any of the color channels, the minimum value of which is the dark pixel. These dark pixels help in the transmission of haze. This is one of the most commonly used methods for dehazing a hazy input image [15], but it cannot remove haze from an image similar to airlight, such as any white object or the headlight of a car. In Anisotropic Diffusion [36], the haze can be removed from any part (lines, edges, etc.) of the image, though the result has very poor visibility.

In this chapter, an efficient single image-based dehazing method is proposed, using Contrast Limited Adaptive Histogram Equalization (CLAHE) and Guided Filter (GF) to remove haze from a hazy image. Histogram Equalization is one of the most commonly used image enhancement techniques, which can only perform well if there is even pixel distribution in the image. Adaptive Histogram Equalization (AHE) can overcome this problem. However, again, the concern with this method is that it tends to over-amplify the contrast of the image. Therefore, CLAHE, a variant of AHE, is introduced. The process is similar to AHE but uses a normalized clip limit of range [0, 1] to reduce over-amplification of noise and contrast. Although it can reduce the problem of noise over-amplification, some information may still be lost because of clipping the histogram. In order to get back the lost information from the CLAHE output image, an edge-preserving GF is applied to the resultant image.

The chapter is organized as follows. Section 12.2 presents the literature survey; Section 12.3 covers the background study. The proposed methodology is given in Section 12.4. Dataset collection and analysis is given in Section 12.5. Image quality assessment criteria is covered in Section 12.6. In Section 12.7, different experimental results along with a discussion are presented. The conclusion and future scope are drawn in Section 12.8.

12.2 LITERATURE SURVEY

In [20], an atmospheric scattering model was established by considering that atmospheric particles become scattered under different weather conditions. After that, various dehazing methods have been proposed to date, based on multiple images, like polarization-based dehazing methods [29, 30], weather-condition-based methods [21, 22, 25], and methods based on Dark Channel Prior.

Dehazing techniques based on physics [15, 34] have also been developed where the image contrast can be restored without using the estimated weather information. But it can only work with some known scene depth. One further multiple image-based dehazing technique is the depth map based method [14, 23], which works on depth information from the user input to dehaze an image. The problem with this method is that it is not automatic but needs user interaction. Another effective method to dehaze an image is the wavelet-based fusion method [8], which can remove the blur effect from the image.

The only limitation with the multiple image-based dehazing method is that it requires more than one image of the same scene, which may be very difficult to get in many scenarios, as is the case for any moving object. As a result, a single image-based dehazing method has gained in popularity as it requires only one input image and, thus, solves the problem faced by the multiple image-based dehazing method.

Paper [40] proposed a defogging method using Contrast Limited AHE. In this method an RGB image is converted into an HSI color model and then CLAHE is used to process the intensity component of the image. The disadvantage of the proposed method is that some of the image information may get lost and also the contrast of the input image may get disturbed. Some advancement techniques of image processing-based applications are described in [5, 17, 28, 32].

One of the most attractive dehazing methods discussed by Jackson et al. in [11] is based on DCP [2], where it is assumed that most of the local patches in a haze-free image contain some pixels, which have very low intensities in any of the color channels (RGB). By using DCP, the transparency or the thickness of the haze can be obtained and thus the quality of the hazy image can be improved. However, the DCP approach has two main limitations: First, it is unable to remove haze from objects which are similar to airlight, like snowy ground or car headlights, and second, the processing time is high for real-time applications. Another method of calculating atmospheric light was proposed by Chaudhary et al. [1]. In this paper, atmospheric light is calculated by using an unbiased bilateral filter with trigonometric range kernels in constant time. Cheng et al. [2], proposed an airlight estimation method by using Gaussian distribution, which has less time-complexity and is more specific than other methods of airlight estimation. DCP, along with a Gaussian filter, is used for dehazing purposes in [19]. Although the proposed method is quite fast and works well on images having uneven haze, it lacks in efficiency in preserving edges of the resultant dehazed image.

A novel method is proposed in [16], by combining both DCP and Bright Channel prior (BCP), in which the airlight is achieved by using the average value of the DCP and BCP. The proposed algorithm effectively achieves a good result in clearing the image, but, as DCP and BCP are a kind of statistic, it may not give the expected result in some particular images. Another improved dehazing method based on the Non-symmetry and Anti-packing Model (NAM) is mentioned in [43] where the airlight is estimated by the above mentioned method and then the scene transmission is estimated by combining boundary constraints and contextual regularization. The approach is effective in reducing the artifacts and gives a high quality image. However, it is a quite time consuming method.

The result of DCP can be improved by using Improved Dark Channel Prior (IDCP) [37], proposed by Ullah et al. Here, DCP with a bilateral filter is used as an improvement in the method. This bilateral filter is used for the purpose of smoothing the image, but it reduces the pixel color saturation and, as a result, the image seems a bit washed out.

A dehazing method based on a GF technique was implemented by He et al. [8]. The proposed method is effective not only in dehazing an image, but also in other fields, like image feathering and edge smoothing. A fast haze removal algorithm was proposed by Sun et al. [33], which increases the contrast of the image and also has a high computation speed.

A fast real-time image dehazing method is presented in [15] which depends on a color cube constraint. This is a simple method of recovering a hazy image by estimating a transmission map using Variation of Distance (VoD) in a prior step. The result obtained is much better as compared to several other methods, and is quite visually appealing. Another fusion-based method to remove haze from an image is described in [16]. In this method, the authors combine a visible image and a near-infrared (NIR) image of the same scene, which creates a natural and clear dehazed image, but fails in removing the haze completely from the image. Recently, a dehazing method for removing the distorted color and enhancing the contrast of a moving object is discussed in [38]. The above method is capable of removing haze from the degraded image that has an uneven haze density.

12.3 BACKGROUND STUDY

12.3.1 WHITE BALANCE (WB)

In the initial stage, WB is applied to the input hazy image. This white balancing helps in identifying or adjusting the actual color of the input image with reference to white, to make the image look more natural. Hazy images generally have a higher color temperature, that is, a colder light (more blue). Because of this, color casts (the tints of a particular unwanted color) may occur which affect the image quality. Therefore, WB correction is applied in the input hazy image for color correction. The pseudocode for the white-balancing procedure is given in Algorithm 12.1.

Algorithm 12.1   White Balance

Input: *Hazy Image, (Img)*
Output: *White Balance Image*
   procedure WHITE_BALANCE(*Img*)
      $[M, N \ Scale] \leftarrow Size(Img)$
      if *Scale* > 1 **then**
         $R\_avg \leftarrow mean(mean(Img(:, :, 1)))$
         $G\_avg \leftarrow mean(mean(Img(:, :, 2)))$
         $B\_avg \leftarrow mean(mean(Img(:, :, 3)))$ ▷ R_avg, G_avg and B_avg are
the average R, G and B color components
      end if
      $RGB\_avg \leftarrow [R\_avg, G\_avg, B\_avg]$
      $Gray\_Value \leftarrow (R\_avg + G\_avg + B\_avg) / 3$
      $Scale\_Value \leftarrow Gray\_Value / RGB\_avg$
      $White\_Balanced(:, :, 1) \leftarrow scaleValue(1) \times Img(:, :, 1)$
      $White\_Balanced(:, :, 2) \leftarrow scaleValue(1) \times Img(:, :, 2)$
      $White\_Balanced(:, :, 3) \leftarrow scaleValue(1) \times Img(:, :, 3)$
   end procedure

12.3.2 CLAHE

Histogram equalization is one of the simplest methods for enhancing an image. It helps in improving the quality of the image without knowing the actual cause of degradation. But it performs well only if the pixel value distribution across the image is identical. CLAHE [26] is the modified version of AHE [13], proposed to limit the problems faced by the latter. CLAHE works in same manner as AHE, the only difference is in the calculation of the clip limit, which specifies the contrast enhancement limit. The higher the value of the clip limit, the more the contrast of the image. This clip limit is based on the normalized histogram, including the size of the neighborhood region.

Initially, the image is divided into different non-overlapping blocks of equal sizes. Next, a histogram for each region is calculated and a desired clip limit is obtained to clip the histograms. The histograms are redistributed such that their heights do not go above the clip limit. This redistribution procedure may need several iterations depending on the clipping factor. Lastly, all the neighboring tiles are merged using bilinear interpolation and their gray scale values are changed according to the modified histograms. The pseudo-code for the CLAHE procedure is given in Algorithm 12.2.

Here, $N_{average}$ average number of pixels N_{levels} is the number of gray levels in contextual regions. NrX is number of pixels in X dimension of contextual region. NrY is the number of pixels in Y dimension contextual region. N_{CLIP} is actual clip limit. N_{clip} is the normalized clip limit in the range [0,1]. N_{total} is the total number of clipped pixels. $N_{avggray}$ is the average number of pixels to be distributed. N_{remain} is the number of pixels to be distributed. $H(i)$ is the original histogram of each region at the ith gray level. $H_{clip}(i)$ is the clipped histogram of each region at the ith gray level.

12.3.3 GF

GF uses the property of another image, for example, the guidance image, to carry out the filtering process. GF can be helpful in many ways, such as being helpful in eliminating false edges and staircase effects from the image, as well as being used in image smoothing and noise reduction from the image. Mathematically, GF can be expressed as:

$$q_i = a_k I_i + b_k, \forall i \in \omega_k \tag{12.3}$$

where the guidance image is represented as I, the input image is represented as p, the output image can be defined as q, ω_k is the size of the window centering at pixel k, and (a_k, b_k) are the constants. The value of a_k and b_k can be calculated as:

$$a_k = \frac{\frac{1}{|\omega|} \sum_{i \in \omega_k} I_i p_i - \mu_k \Gamma_k}{\sigma_k^2 + \varepsilon} \tag{12.4}$$

$$b_k = \Gamma_k - a_k \mu_k \tag{12.5}$$

Algorithm 12.2   CLAHE

Input: $Input_{img}$
Output: $Enhanced_{img}$

1. Divide the original intensity image into non-overlapping contextual regions. The total number of image tiles is equal to $M \times N$.
2. Calculating the histogram of each contextual region according to gray levels present in the array image.
3. Calculating the contrast limited adaptive histogram of the contextual region by the Clipped Limit value as

$$N_{average} = (NrX \times NrY)/N_{levels}$$
$$N_{CLIP} = N_{clip} \times N_{gray}$$

The average number of pixels redistributed is given by

$$N_{avggray} = N_{total}/N_{levels}$$

4. To clip the pixels
    procedure CLIP
      if $H(i) > N_{CLIP}$ **then**
        $H_{clip} = N_{CLIP}$
      else
        if $(H(i) + N_{avggray}) > N_{CLIP}$ **then**
          $H_{clip} = N_{CLIP}$
        else
          $H_{clip}(i) = N_{CLIP} + H(i)$
        end if
      end if
    end procedure
5. Redistribute the remaining pixels until they have all been distributed. The step of redistributed pixels is given by N_{levels}/N_{remain}
6. Calculate the new gray level assignment of pixels within a submatrix contextual region by using interpolation.

---

where ω is the number of pixels in the window, μ_k and σ_k^2 is the mean and variance of I respectively, ω_k, ε is the regularization parameter to avoid maximizing the value of a_k, and p_i is the mean of p in window ω_k.

A linear model can be applied to every local window in the whole image. But, a pixel is included in every patch window ω_k that has this pixel. An easy method is to find the mean of every possible value of the pixel of the output image q_i which can be shown as:

$$q_i = \frac{1}{|\omega_i|} \sum_{k \in \omega_i} a_k I_i + b_k \tag{12.6}$$

$$= \bar{a}_i I_i + \bar{b}_i \tag{12.7}$$

where $\bar{a}_i$ and $\bar{b}_i$ are the mean in the patch of a_k and b_k respectively.

12.4 PROPOSED METHODOLOGY

The proposed algorithm is explained in the following steps:

1. Initially, a hazy image is taken as an input image.
2. The white balancing procedure, as given in Algorithm 12.1, is applied to the hazy image so as to adjust the color of the input image with reference to a white color, so as to make the image look more natural.
3. To improve the visibility of the hazy image, the CLAHE procedure, as given in Algorithm 12.2, is applied to the white balanced image. In this procedure, initially, the image is divided into non-overlapping blocks, and a histogram of each block is calculated. After that a clip limit is chosen to clip the parts of the histogram which are exceeding the limit.
4. Finally, the histogram of every region is combined with bilinear interpolation and image gray-scale values are changed according to the new histograms.
5. Due to clipping of the histogram in CLAHE, some of the information of the image may get lost. So, to get back that information, GF is used in this step. For GF, a guidance image (I) is considered, using which the filtering process is done.
6. The mean of (I), the mean of the CLAHE output image (p), and the mean of ($I \times p$) is calculated.
7. Co-variance of (I, p) is calculated as:

$$cov\_Ip = mean\_Ip - mean\_I.*mean\_p \qquad (12.8)$$

8. The variance of (I) is obtained to get linear coefficients a and b using:

$$var\_I = mean\_II - mean\_I.*mean\_I \qquad (12.9)$$

9. The values of a and b are important to get the final result, which can be obtained by:

$$a = (cov\_Ip)./(var\_I + \epsilon) \qquad (12.10)$$

$$b = mean\_p - a*mean\_I \qquad (12.11)$$

where ϵ is the regularizing parameter.
10. And finally, the mean of a and b is calculated and the final output of the image, i.e., q, is obtained as:

$$q = mean\_a.*I + mean\_b \qquad (12.12)$$

12.5 DATASET COLLECTION AND ANALYSIS

In this chapter, the Haze Realistic Dataset or HazeRD [42] is used for evaluating the proposed algorithm. HazeRD consists of 15 scenes along with their haze-free RGB images of their corresponding depth maps. Five distinct weather conditions are imitated into each of the scenes, which range in ascending order of their visual density,

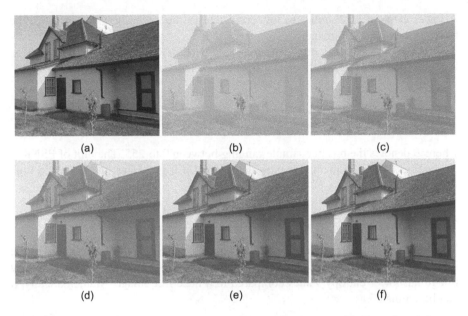

(a)   (b)   (c)

(d)   (e)   (f)

FIGURE 12.1 Sample of HazeRD having different visual ranges. (a) Ground truth image, (b) image with visual range 50 m, (c) image with visual range 100 m, (d) image with visual range 200 m, (e) image with visual range 500 m, (f) image with visual range 1000 m.

labelled as 50, 100, 200, 500, and 1000. HazeRD is preferred over other dehazing datasets as it provides outdoor scenes, whereas other datasets focus basically on indoor scenes. Another advantage is that this dataset produces haze using real-life parameters, which are more realistic than the prior datasets, using either indoor images or synthetically produced images. In HazeRD, a Matlab function and a demo script is available as well, to create a hazy image with different parameters of haze. It also has the option of noise (by default it is off) for avoiding some unnatural uniformity in the airlight regions. Figure 12.1 shows an example of the dataset image.

12.6 IMAGE QUALITY ASSESSMENT CRITERIA

12.6.1 Peak Signal-to-Noise Ratio and Mean Squared Error

Peak Signal-to-Noise Ratio (PSNR) and Mean Squared Error (MSE) [17, 18, 27, 41] are the two most used error matrices for comparing the quality of the dehazed image. The cumulative squared error among improved dehazed images and the original image is represented by MSE, which can mathematically be represented as:

$$\text{MSE} = \frac{1}{[M \times N]} \sum_{M,N} \left[I_1(m,n) - I_2(m,n) \right] \tag{12.13}$$

Here, $I_1(m,n)$ and $I_2(m,n)$ are the hazy image and the improved image respectively, $M \times N$ is the size of the image, and m, n signifies the x, y location of the image pixel.

A lower value of MSE indicates low error, that is, the lower the value of MSE, the better is the quality of image.

A measure of the peak error is represented by PSNR, which can be described as the reciprocal of MSE. Mathematically, it can be represented as:

$$PSNR = 10 \times \log_{10} \frac{\left(2^B - 1\right)^2}{MSE} \tag{12.14}$$

where B denotes the bits per sample ranging between 0 to 255. The unit of PSNR is the decibel (dB). The method is considered good if it generates low MSE and high PSNR values.

12.6.2 ENTROPY

The entropy of an image identifies its texture and measures the randomness of the image as well. Low entropy specifies that the region is homogeneous, thus a hazy image has lower entropy values compared to a haze free image. Mathematically, it can be estimated as:

$$H = -\sum_i p_i \log_2 p_i \tag{12.15}$$

where p_i indicates the probability of occurrence of one gray level intensity.

12.6.3 STRUCTURAL SIMILARITY INDEX

This index helps in measuring image degradation caused by various processing. Its calculation requires a reference image. Mathematically,

$$SSIM(x, y) = \frac{\left(2\mu_x\mu_y + c_1\right)\left(2\sigma_{xy} + c_2\right)}{\left(\mu_x^2 + \mu_y^2 + c_1\right)\left(\sigma_x^2 + \sigma_y^2 + c_2\right)} \tag{12.16}$$

Here, μ_x and μ_y are the mean of x and y respectively, $\sigma_x^2 \sigma_y^2$ are the variance of x and y respectively, and σ_{xy} is the co-variance of x, y. c_1 and c_2 are the variables used for stabilizing the division.

12.6.4 CONTRAST GAIN

Contrast gain is the mean difference between the contrast of the enhanced image and the original image (hazy image) [6]. It can be calculated by:

$$C_{gain} = C_j - C_i \tag{12.17}$$

where C_j is the mean contrast of the output enhanced image and C_i is the contrast of the original hazy image.

The higher the value of the contrast of the enhanced image, the better is the result for the dehazing algorithm.

12.7 EXPERIMENTAL RESULTS AND DISCUSSION

Similarly, in Figure 12.3, haze level 100 is considered. In Figure 12.3(c), the output of the white balanced image is given. After applying CLAHE to it, the result obtained seems to be not very appealing, as shown in Figure 12.3(d). To improve the result, GF has been used and its result is shown in Figure 12.3(e). It is observed that the result is quite successful in removing the haze from the background as well.

The experiment has been performed and tested on different hazy images collected from the data set HazeRD. For Figure 12.2, haze level 50 has been considered. Figure 12.2(b) shows that the image is quite hazy. White balancing is performed to adjust the color of the image and the output of the white balanced image is shown in Figure 12.2(c). After applying CLAHE, in Figure 12.2(d), not much visual improvement is observed in the image. So the image is tested by applying GF to it in Figure 12.2(e), where it is observed that, although the visibility of the hazy image has increased, the haze has not been removed totally from the background. Considering haze level 200 for Figure 12.4, it is observed that the result obtained after applying CLAHE, shown in Figure 12.4(d), is not improved much as compared to the resultant image of white balance in Figure 12.4(c). So, to improve the result, GF is applied and the final resultant image is shown in Figure 12.4(e), where haze is seen to be removed and the result obtained is also very appealing. The contrast of the image is also improved as compared to the earlier obtained results.

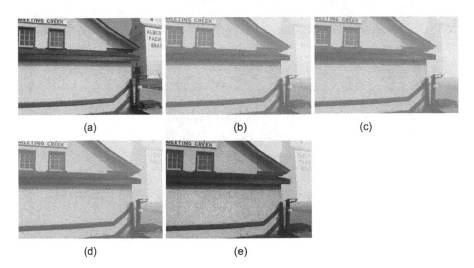

(a)   (b)   (c)

(d)   (e)

FIGURE 12.2 Experimental results on hazy Image 50: (a) Ground truth image, (b) Input image, (c) Image after white balancing, (d) Image after applying CLAHE, (e) Image after applying CLAHE+Guided.

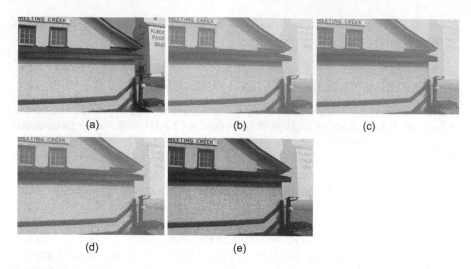

FIGURE 12.3 Experimental results on hazy Image 100: (a) Ground truth image, (b) Input image, (c) Image after white balancing, (d) Image after applying CLAHE, (e) Image after applying CLAHE+Guided.

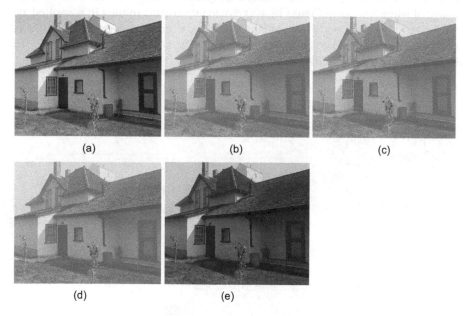

FIGURE 12.4 Experimental results on Hazy Image 200: (a) Ground truth image, (b) Input image, (c) Image after white balancing, (d) Image after applying CLAHE, (e) Image after applying CLAHE+Guided.

FIGURE 12.5   Experimental results on hazy Image 500: (a) Ground truth image, (b) Input image, (c) Image after white balancing, (d) Image after applying CLAHE, (e) Image after applying CLAHE+Guided.

Finally, for Figures 12.5 and 12.6, haze level 500 is considered and the result obtained after applying the proposed algorithm is shown in Figures 12.5(e) and 12.6(e). It can be observed that, the result obtained in both the cases is quite successful in removing haze completely from the image.

So, from the experimental results, it is clear that the proposed algorithm gives a successful result in removing haze from the images having different haze levels.

In Tables 12.1, 12.2, 12.3, and 12.4, the entropy of images, having different haze levels, is calculated individually for each case, that is, for the input hazy image, the WB image, the CLAHE output image, and the proposed algorithm. From the result obtained, it is clearly noticeable that, in every case, the entropy has increased gradually, which further satisfies the definition of entropy as discussed in Section 12.6.

In this chapter, comparison of the contrast gain of the white balanced image, the CLAHE image, and the proposed algorithm is given in Table 12.5, where it can be noticed that the contrast gain is maximum in the case of the proposed algorithm. Similarly, in Table 12.6, a Structural Similarity Index (SSIM) is given and, by comparing the results, it can be observed that the proposed method is better than the individual results obtained from the white balance image and the CLAHE image.

In Tables 12.7, 12.8, 12.9, and 12.10, PSNR and MSE are calculated for each case, that is, for the white balanced image, the CLAHE output image, and the proposed algorithm. It is observed that the values of PSNR obtained are increasing and the values of MSE are decreasing gradually for all the images having different haze levels. Thus it justifies the definition of PSNR and MSE explained in Section 12.6.

For selecting the value of the clip limit, experiments have been performed by taking different values ranging from 0.001 to 0.02. In Figure 12.7(a), the change in

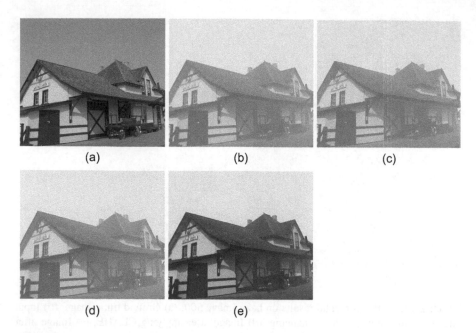

FIGURE 12.6 Experimental results on hazy Image 500: (a) Ground truth image, (b) Input image, (c) Image after white balancing, (d) Image after applying CLAHE, (e) Image after applying CLAHE+Guided.

TABLE 12.1
Entropy for the Different Images (Haze Level 50)

Images	Input_Img	WB	CLAHE	Proposed_Algo
8905_50	4.5328	4.6160	4.1229	5.5894
7033_50	4.7523	4.7523	5.0321	6.1805
8612_50	4.4142	5.0283	4.9598	5.6666
8411_50	4.7069	4.7069	4.9769	5.7104
8602_50	1.0156	1.0156	1.0485	1.3057
9562_50	4.0313	4.3796	4.4768	5.2476
8583_50	6.4290	6.5397	6.6828	7.4777
8895_50	3.5613	3.5613	3.5779	4.1139
7460_50	4.6813	5.0491	4.9275	5.9397
8503_50	5.6430	5.7572	5.8784	6.6873

PSNR values have been plotted against different clip limits, and it is observed that the PSNR values have increased up to 0.004 and, after that, have gradually decreased. It is also observed that from 0.01 the values of PSNR remain almost constant, for all the different haze levels.

Similarly, for Figure 12.7(b), experiments were performed by taking MSE values and different clip limits, and the change in results is plotted in the graph. It can clearly be noted that, considering all the haze levels, MSE values are changing from 0.001 to 0.004 and then increasing; from 0.01 the values are nearly constant. In the

TABLE 12.2
Entropy for the Different Images (Haze Level 100)

Images	Input_Img	WB	CLAHE	Proposed_Algo
8905_100	5.4160	5.5104	5.5906	6.3851
7033_100	5.0342	5.4662	5.5616	6.3144
8612_100	4.7288	5.0960	5.4716	5.9133
8411_100	5.3070	5.4953	5.6993	6.0931
8602_100	1.5804	1.5804	1.8197	2.5705
9562_100	4.4190	4.7700	5.2135	5.7217
8583_100	6.8261	6.8399	7.0181	7.5408
8895_100	3.8872	4.3395	4.3820	4.7204
7460_100	5.3353	5.5628	5.8535	6.4265
8503_100	6.0485	6.2272	6.3889	6.7497

TABLE 12.3
Entropy for the Different Images (Haze Level 200)

Images	Input_Img	WB	CLAHE	Proposed_Algo
8905_200	5.9866	6.1446	6.2828	6.6998
7033_200	5.1729	5.6113	5.7430	6.2128
8612_200	4.9321	5.5544	5.7897	6.0988
8411_200	5.7225	6.0526	6.2519	6.5294
8602_200	1.8171	1.8171	2.2702	3.5205
9562_200	4.7325	5.3207	5.6718	6.0418
8583_200	7.0706	7.0584	7.2009	7.6185
8895_200	4.0654	4.8206	4.6535	4.9594
7460_200	5.5725	5.9640	6.2490	6.7158
8503_200	6.3318	6.4996	6.7348	6.8685

TABLE 12.4
Entropy for the Different Images (Haze Level 500)

Images	Input_Img	WB	CLAHE	Proposed_Algo
8905_500	6.4760	6.6434	6.8270	6.9286
7033_500	5.3028	5.7274	5.9648	6.3246
8612_500	5.1043	5.4662	6.0523	6.2022
8411_500	6.0298	6.3569	6.7168	6.6860
8602_500	1.9799	1.9799	2.5831	3.9649
9562_500	5.0703	5.6726	6.1249	6.4400
8583_500	7.3283	7.2990	7.4185	7.6981
8895_500	4.1707	4.9281	5.0623	5.1402
7460_500	5.7061	6.0856	6.4953	6.7344
8503_500	6.6007	6.7825	7.0328	6.9597

TABLE 12.5
Contrast Gain for Different Images

Images	WB	CLAHE	Proposed_Algo
8905_50	0.0094	0.0175	0.2147
8583_50	0.0358	0.0746	0.5171
8612_100	0.0098	0.0497	0.4783
8583_100	0.6488	0.6642	0.8107
8602_200	0.2600	0.2619	0.4303
8503_200	0.0238	0.0679	0.5055
7460_500	0.0308	0.0437	0.3854
8612_500	0.0192	0.0471	0.4381

TABLE 12.6
SSIM for Different Images

Images	WB	CLAHE	Proposed_Algo
8905_50	0.3977	0.4325	0.5913
8583_50	0.5753	0.5872	0.7712
8612_100	0.3725	0.3807	0.4922
8583_100	0.6488	0.6642	0.8107
8602_200	0.2600	0.2619	0.3529
8503_200	0.6596	0.6792	0.7476
7460_500	0.5933	0.6110	0.6772
8612_500	0.5770	0.5960	0.5922

TABLE 12.7
PSNR and MSE for Different Images (Haze Level 50)

	WB		CLAHE		Proposed_Algo	
Images	PSNR	MSE	PSNR	MSE	PSNR	MSE
8905_50	60.3165	0.1817	60.3714	0.1793	65.8521	0.0615
7033_50	61.4570	0.1397	61.5453	0.1369	67.2987	0.0364
8612_50	60.6204	0.1702	60.7435	0.1655	63.4328	0.0910
8411_50	62.2395	0.1142	62.4212	0.1119	67.4410	0.0354
8602_50	62.4931	0.1178	62.5186	0.1172	63.1124	0.1036
9562_50	60.8137	0.1622	60.9017	0.1590	63.1684	0.0954
8583_50	64.7303	0.0661	65.3364	0.0575	68.1197	0.0301
8895_50	61.0609	0.1528	61.1456	0.1501	63.5542	0.0866
7460_50	60.0017	0.1960	60.0638	0.1932	61.8371	0.1284
8503_50	62.2970	0.1156	62.5149	0.1100	71.0371	0.0156

same way, considering Figure 12.7(c), (d), and (e), the values of SSIM, entropy, and contrast gain are increasing up to 0.004 and then decreasing, which is clearly visible from the given graphs.

So, considering all the results, discussed in the above section, a clip limit 0.004 is assumed to be the ideal limit for the proposed algorithm.

TABLE 12.8
PSNR and MSE for Different Images (Haze Level 100)

	WB		CLAHE		Proposed_Algo	
Images	PSNR	MSE	PSNR	MSE	PSNR	MSE
8905_100	62.4693	0.1108	62.6827	0.1055	72.6006	0.0108
7033_100	64.0733	0.0766	64.2759	0.0731	72.5439	0.0109
8612_100	62.2252	0.1185	62.4898	0.1117	65.1923	0.0629
8411_100	64.7772	0.0651	65.0512	0.0611	71.0179	0.0155
8602_100	62.7623	0.1116	62.8182	0.1055	63.8557	0.0893
9562_100	62.4542	0.1118	62.7117	0.1055	65.8580	0.0538
8583_100	67.0228	0.0391	67.9396	0.0316	69.2701	0.0231
8895_100	62.8090	0.1026	62.9701	0.0989	66.6015	0.0438
7460_100	61.3515	0.1437	61.5766	0.1365	64.8669	0.0641
8503_100	65.1706	0.0599	65.8429	0.0514	73.6977	0.0084

TABLE 12.9
PSNR and MSE for Different Images (Haze Level 200)

	WB		CLAHE		Proposed_Algo	
Images	PSNR	MSE	PSNR	MSE	PSNR	MSE
8905_200	65.4788	0.0555	66.2153	0.0469	74.3319	0.0072
7033_200	67.3685	0.0359	67.8051	0.0325	74.3695	0.0075
8612_200	63.8295	0.0832	64.2817	0.0755	66.0236	0.0534
8411_200	67.8332	0.0323	68.6579	0.0267	70.9779	0.0156
8602_200	63.1058	0.1043	63.1808	0.1028	65.1617	0.0680
9562_200	64.4249	0.0721	64.9524	0.0643	68.2702	0.0347
8583_200	69.6355	0.0215	69.2761	0.0131	72.8459	0.0016
8895_200	64.9485	0.0632	65.2555	0.0590	69.6534	0.0237
7460_200	63.3218	0.0915	63.8360	0.0814	69.4296	0.0228
8503_200	68.6587	0.0271	70.1902	0.0190	73.2222	0.0189

TABLE 12.10
PSNR and MSE for Different Images (Haze Level 500)

	WB		CLAHE		Proposed_Algo	
Images	PSNR	MSE	PSNR	MSE	PSNR	MSE
8905_500	67.3462	0.0261	69.054	0.0108	75.7431	0.0061
7033_500	69.4035	0.0358	70.5552	0.0186	76.3774	0.0045
8612_500	65.3550	0.0603	65.7849	0.0554	68.4110	0.0059
8411_500	70.4891	0.0221	72.2382	0.0175	75.2609	0.0067
8602_500	63.4477	0.0977	63.5267	0.0962	66.1265	0.0558
9562_500	66.8301	0.0438	67.5183	0.0386	67.5795	0.0386
8583_500	70.2149	0.0205	72.4373	0.0117	75.3941	0.0082
8895_500	67.6465	0.0350	68.2060	0.0312	70.2276	0.0205
7460_500	62.2712	0.0468	67.2298	0.0377	69.3260	0.0235
8503_500	71.8090	0.0287	74.8268	0.0178	78.5556	0.0044

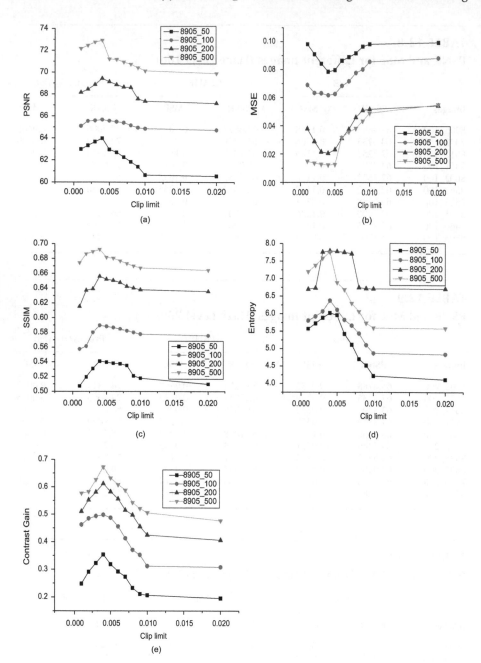

FIGURE 12.7   Experiment results at different clip Limits: (a) PSNR vs clip limit, (b) MSE vs clip limit, (c) SSIM vs clip limit, (d) Entropy vs clip limit, (e) Contrast gain vs clip limit.

12.8 CONCLUSION AND FUTURE SCOPE

The dehazing method presented in this chapter combines two methods together, that is, CLAHE and GF. But before applying the proposed method, white balancing is carried out to the hazy input image to make it look more natural. After white balancing, CLAHE is applied to the image for enhancement, but as a result of clipping the histogram in CLAHE, some information is lost, for which GF was used to retrieve it.

For selecting the clip limit value, different experiments were performed on images having different haze levels. By taking different clip limits, ranging from 0.001 to 0.02, performance criteria like PSNR, MSE, SSIM, entropy, and contrast gain were analyzed and the results obtained were plotted in the corresponding graphs. By observing the graphs, it is noticed that at clip limit 0.004, the results obtained in each case are quite good. So, for experimental purposes, a clip limit of 0.004 is considered.

From the experimental results performed on the dataset, it can be concluded that the proposed algorithm has quite good results in removing haze. Also by analyzing the performance criteria of entropy, PSNR, MSE, and SSIM, it can be said that the proposed algorithm is effective in removing haze from any image.

In future, we will try to improve the proposed algorithm to dehaze a hazy image in the case of a very low contrast image.

REFERENCES

1. Chaudhury, K.N., Sage, D., Unser, M.: Fast $o(1)$ bilateral filtering using trigonometric range kernels. *IEEE Transactions on Image Processing* 20(12), 3376–3382 (2011).
2. Cheng, F.C., Cheng, C.C., Lin, P.H., Huang, S.C.: A hierarchical airlight estimation method for image fog removal. *Engineering Applications of Artificial Intelligence* 43, 27–34 (2015).
3. Das, H., Jena, A.K., Badajena, J.C., Pradhan, C., Barik, R.: Resource allocation in cooperative cloud environments. In: *Progress in Computing, Analytics and Networking*, pp. 825–841. Singapore: Springer (2018).
4. Das, H., Naik, B., Behera, H.: An experimental analysis of machine learning classification algorithms on biomedical data. In: *Proceedings of the 2nd International Conference on Communication, Devices and Computing*, pp. 525–539. Singapore: Springer (2020).
5. Economopoulos, T., Asvestas, P., Matsopoulos, G.: Contrast enhancement of images using partitioned iterated function systems. *Image and Vision Computing* 28(1), 45–54 (2010). https://doi.org/https://doi.org/10.1016/j.imavis.2009.04.011, http://www.sciencedirect.com/science/article/pii/S026288560900081X
6. Fattal, R.: Single image dehazing. *ACM Transactions on Graphics* 27(3), 72:1–72:9 (August 2008). https://doi.org/10.1145/1360612.1360671, http://doi.acm.org/10.1145/1360612.1360671
7. Grewe, L.L., Brooks, R.R.: Atmospheric attenuation reduction through multisensor fusion. In: *Sensor Fusion: Architectures, Algorithms, and Applications II*, vol. 3376, pp. 102–110. Bellingham, WA: International Society for Optics and Photonics (1998).
8. He, K., Sun, J., Tang, X.: Single image haze removal using dark channel prior. *IEEE Transactions on Pattern Analysis and Machine Intelligence* 33(12), 2341–2353 (December 2011). https://doi.org/10.1109/TPAMI.2010.168
9. He, K., Sun, J., Tang, X.: Guided image filtering. *IEEE Transactions on Pattern Analysis and Machine Intelligence* 35(6), 1397–1409 (June 2013). https://doi.org/10.1109/TPAMI.2012.213

10. Tan, K., Oakley, J.P.: Enhancement of color images in poor visibility conditions. In: *Proceedings 2000 International Conference on Image Processing* (Cat. No.00CH37101), Vancouver, Canada, vol. 2, pp. 788–791 (September 2000). https://doi.org/10.1109/ICIP.2000.899827

11. Jackson, J., Ariyo, O., Acheampong, K., Boakye, M., Frimpong, E., Ashalley, E., Rao, Y.: Hybrid single image dehazing with bright channel and dark channel priors. In: *2017 2nd International Conference on Image, Vision and Computing (ICIVC)*, Chengdu, China, pp. 381–385 (June 2017). https://doi.org/10.1109/ICIVC.2017.7984582

12. Jha, R.K., Soni, B., Aizawa, K.: Logo extraction from audio signals by utilization of internal noise. *IETE Journal of Research* 59(3), 270–279 (2013).

13. Ketcham, D.J.: Real-time image enhancement techniques. In: *Image Processing*, vol. *74*, pp. 120–125 (1976).

14. Kopf, J., Neubert, B., Chen, B., Cohen, M., Cohen-Or, D., Deussen, O., Uyttendaele, M., Lischinski, D.: Deep photo: Model-based photograph enhancement and viewing. *ACM Transactons on Graphics* 27(5), 116:1–116:10 (December 2008). https://doi.org/10.1145/1409060.1409069, http://doi.acm.org/10.1145/1409060.1409069

15. Kponou, E.A., Wang, Z., Wei, P.: Efficient real-time single image dehazing based on color cube constraint. In: *2017 IEEE 2nd International Conference on Signal and Image Processing (ICSIP)*, Singapore, pp. 106–110 (August 2017). https://doi.org/10.1109/SIPROCESS.2017.8124515

16. Kudo, Y., Kubota, A.: Image dehazing method by fusing weighted near-infrared image. In: *2018 International Workshop on Advanced Image Technology (IWAIT)*, Chiang Mai, Thailand, pp. 1–2 (January 2018). https://doi.org/10.1109/IWAIT.2018.8369744

17. Lin, M.H., Hu, Y.C., Chang, C.C.: Both color and gray scale secret images hiding in a color image. *International Journal of Pattern Recognition and Artificial Intelligence* 16(06), 697–713 (2002).

18. Liu, T., ding Qiu, Z.: A DWT-based color image steganography scheme. In: *6th International Conference on Signal Processing*, Beijing, China, vol. 2, pp. 1568–1571 (August 2002). https://doi.org/10.1109/ICOSP.2002.1180096

19. Long, J., Shi, Z., Tang, W., Zhang, C.: Single remote sensing image dehazing. *IEEE Geoscience and Remote Sensing Letters* 11(1), 59–63 (January 2014). https://doi.org/10.1109/LGRS.2013.2245857

20. McCartney, E.J.: *Optics of the Atmosphere: Scattering by Molecules and Articles*. New York: John Wiley & Sons, Inc., 421pp. (1976).

21. Narasimhan, S.G., Nayar, S.K.: Chromatic framework for vision in bad weather. In: *Proceedings IEEE Conference on Computer Vision and Pattern Recognition. CVPR 2000* (Cat. No.PR00662), Hilton Head Island, SC, vol. 1, pp. 598–605 (June 2000). https://doi.org/10.1109/CVPR.2000.855874

22. Narasimhan, S.G., Nayar, S.K.: Contrast restoration of weather degraded images. *IEEE Transactions on Pattern Analysis and Machine Intelligence* 25(6), 713–724 (June 2003). https://doi.org/10.1109/TPAMI.2003.1201821

23. Narasimhan, S.G., Nayar, S.: Interactive deweathering of an image using physical models. In: *IEEE Workshop on Color and Photometric Methods in Computer Vision, In Conjunction with ICCV*, Nice, France (October 2003).

24. Narasimhan, S.G., Nayar, S.K.: Vision and the atmosphere. *International Journal of Computer Vision* 48(3), 233–254 (2002)

25. Nayar, S.K., Narasimhan, S.G.: Vision in bad weather. In: *Proceedings of the Seventh IEEE International Conference on Computer Vision*, Kerkyra, Greece, vol. 2, pp. 820–827 (September 1999). https://doi.org/10.1109/ICCV.1999.790306

26. Reza, A.M.: Realization of the contrast limited adaptive histogram equalization (CLAHE) for real-time image enhancement. *VLSI Signal Processing* 38, 35–44 (2004).

27. Rongrong, N., Qiuqi, R.: Embedding information into color images using wavelet. In: *2002 IEEE Region 10 Conference on Computers, Communications, Control and Power Engineering. TENCOM '02. Proceedings*, Beijing, China, vol. 1, pp. 598–601 (October 2002). https://doi.org/10.1109/TENCON.2002.1181346

28. Sahoo, A.K., Pradhan, C., Das, H.: Performance evaluation of different machine learning methods and deep-learning based convolutional neural network for health decision making. In: *Nature Inspired Computing for Data Science*, pp. 201–212. Cham: Springer (2020)

29. Schechner, Y.Y., Narasimhan, S.G., Nayar, S.K.: Instant dehazing of images using polarization. In: *Proceedings of the 2001 IEEE Computer Society Conference on Computer Vision and Pattern Recognition. CVPR 2001*, Kauai, HI, vol. 1, pp. I (December 2001). https://doi.org/10.1109/CVPR.2001.990493

30. Shwartz, S., Namer, E., Schechner, Y.Y.: Blind haze separation. In: *2006 IEEE Computer Society Conference on Computer Vision and Pattern Recognition (CVPR'06)*, New York, vol. 2, pp. 1984–1991 (June 2006). https://doi.org/10.1109/CVPR.2006.71

31. Soni, B., Das, P.K., Thounaojam, D.M.: Keypoints based enhanced multiple copy-move forgeries detection system using density-based spatial clustering of application with noise clustering algorithm. *IET Image Processing* 12(11), 2092–2099 (2018).

32. Soni, B., Das, P.K., Thounaojam, D.M.: Multicmfd: Fast and efficient system for multiple copy-move forgeries detection in image. In: *Proceedings of the 2018 International Conference on Image and Graphics Processing*, Hong Kong, pp. 53–58 (2018).

33. Sun, C.C., Lai, H.C., Sheu, M.H., Huang, Y.H.: Single image fog removal algorithm based on an improved dark channel prior method. In: *2016 International Symposium on Intelligent Signal Processing and Communication Systems (ISPACS)*, Phuket, Thailand, pp. 1–4. IEEE (2016).

34. Tan, K., Oakley, J.P.: Physics-based approach to color image enhancement in poor visibility conditions. *JOSA A* 18(10), 2460–2467 (2001).

35. Tan, R.T.: Visibility in bad weather from a single image. In: *2008 IEEE Conference on Computer Vision and Pattern Recognition*, Anchorage, AK, pp. 1–8 (June 2008). https://doi.org/10.1109/CVPR.2008.4587643

36. Tripathi, A.K., Mukhopadhyay, S.: Single image fog removal using anisotropic diffusion. *IET Image Processing* 6(7), 966–975 (October 2012). https://doi.org/10.1049/iet-ipr.2011.0472

37. Ullah, E., Nawaz, R., Iqbal, J.: Single image haze removal using improved dark channel prior. In: *2013 5th International Conference on Modelling, Identification and Control (ICMIC)*, Cairo, Egypt, pp. 245–248 (August 2013).

38. Wang, P., Fan, Q., Zhang, Y., Bao, F., Zhang, C.: A novel dehazing method for color fidelity and contrast enhancement on mobile devices. *IEEE Transactions on Consumer Electronics* 65(1), 47–56 (February 2019). https://doi.org/10.1109/TCE.2018.2884794

39. Wang, Y., Wu, B.: Improved single image dehazing using dark channel prior. In: *2010 IEEE International Conference on Intelligent Computing and Intelligent Systems*, Xiamen, China, vol. 2, pp. 789–792 (October 2010). https://doi.org/10.1109/ICICISYS.2010.5658614

40. Xu, Z., Liu, X., Ji, N.: Fog removal from color images using contrast limited adaptive histogram equalization. In: *2009 2nd International Congress on Image and Signal Processing*, Tianjin, China, pp. 1–5 (October 2009). https://doi.org/10.1109/CISP.2009.5301485

41. Yu, Y.H., Chang, C.C., Lin, I.C.: A new steganographic method for color and grayscale image hiding. *Computer Vision and Image Understanding* 107(3), 183–194 (2007).

42. Zhang, Y., Ding, L., Sharma, G.: Hazerd: An outdoor scene dataset and benchmark for single image dehazing. In: *2017 IEEE International Conference on Image Processing (ICIP)*, Beijing, China, pp. 3205–3209 (September 2017). https://doi.org/10.1109/ICIP.2017.8296874

43. Zheng, Y., Xie, Z., Cai, C.: Single image dehazing using non-symmetry and anti-packing model based decomposition and contextual regularization. In: *2017 13th International Conference on Natural Computation, Fuzzy Systems and Knowledge Discovery (ICNC-FSKD)*, Guilin, China, pp. 621–625 (July 2017). https://doi.org/10.1109/FSKD.2017.8393342

Index

Milton Keynes UK
Ingram Content Group UK Ltd.
UKHW040107071024
449327UK00019B/866